GEMSTONES

SECOND EDITION

CALLY OLDERSHAW, CHRISTINE WOODWARD AND ROGER HARDING

PUBLISHED BY THE NATURAL HISTORY MUSEUM, LONDON

First printed 2001, reprinted 2002 and 2004

Published by The Natural History Museum,
Cromwell Road, London SW7 5BD

ISBN 0 565 09155 7

A catalogue record for this book is available from the British Library

Edited by Karin Fancett
Designed by David Robinson and Mercer Design
Reproduction and printing by Craft Print, Singapore

Distribution

North America including South America, Central America and the Caribbean
Sterling Publishing Co. Inc.
387 Park Avenue South
New York, NY 10016 – 8810
USA

Canada
Canadian Manda Group
165 Dufferin St.
Toronto
Ontario M6K 3H6
Canada

Australia and New Zealand
CSIRO Publishing
PO Box 1139
Collingwood, Victoria 3066
Australia

UK and the rest of the world
NBN International
Estover Road
Plymouth, Devon, PL6 7PY
UK

Contents

Preface 4
About the authors 4
Is it a gem? 5
Finding gems 7
Properties 12
Passage of light 16
Cutting and polishing 20
Real and fake 22
Jewellery 29
The gemstones 34
Diamond 35
Ruby and Sapphire 38
Emerald and Aquamarine 40
Opal 42
Amethyst and Citrine 44
Garnet 46
Tourmaline 48
Topaz 49
Chrysoberyl 50
Peridot 51
Zircon 52
Spinel 53
Tanzanite 54
Taaffeite 54
Benitoite 55
Iolite (cordierite) 55
Andalusite, Fibrolite (sillimanite) and Kyanite 56
Spodumene 57
Kornerupine 58
Scapolite 58
Diopside and Enstatite 59
Sphene (titanite) 59
Chalcedony and Jasper 60
Jade 62
Turquoise 64
Malachite and Azurite 65
Lapis lazuli 66
Moonstone and Labradorite 68
Blue John 69
Rhodochrosite and Rhodonite 69
Serpentine 70
Ivory and Bone 70
Amber and Jet 71
Pearls, Coral and Shell 72
Further reading and credits 73
Gem deposits of the world 74
Index 76

Preface

Gemstones have been a source of delight and fascination for many thousands of years. The first part of this book explores the nature of gemstones and their special qualities of beauty, rarity and durability, which are determined by the properties of the minerals, and provide the means of identifying gems. Gem mineral formation is introduced, with examples of where they are found and how they are mined. There is also a short section on how gems are identified. Real and fake gemstones are explained, and techniques for their identification are introduced.

The second part of the book gives information about the individual gem species. The main characteristics of the gems, and their chemical and physical properties are described, together with notes on items of particular interest. The authors have had the good fortune to work with one of the world's finest public collections of gemstones held at The Natural History Museum, London.

About the authors

Cally Oldershaw works as the Researcher and Parliamentary Liaison Officer for the Geological Society, London. She is a freelance author and consultant on geoscientific topics. Cally previously worked at The Natural History Museum, London, where she was the Museum Gemmologist responsible for the gemstone collections. Cally is an examiner for the Gemmological Association (GAGTL) and the British examiner for the European exams in gemmology (FEEG). Christine Woodward prepared and produced geological exhibitions at The Natural History Museum, until 1990. She continues to work with the Museum delivering adult education courses relating to mineralogy. Since 1982 she has been an examiner for the GAGTL, and has actively been involved in their educational programme. Roger Harding is the Director of the GAGTL, a professional body providing membership, educational outreach and gem testing services. Prior to this post he was curator of gemstones at The Natural History Museum, from 1985 to 1990.

Is it a gem?

The desire for jewellery and beautiful objects is one that we share with our earliest ancestors. Our rings, lucky charms and royal regalia reflect a tradition of self-adornment, magic and ritual, which extends back many thousands of years.

From ancient times, many materials, both natural and artificial, have been set in jewellery and other precious objects. Over the centuries, however, the term 'gemstone' has come to mean a naturally occurring mineral desirable for its beauty, valuable in its rarity and sufficiently durable to give lasting pleasure.

Beauty

Gems display an almost limitless variety in their beauty. Qualities as diverse as the fiery brilliance of diamond and the soft iridescence of pearl, the bold patterns of agate and pale translucency of jade, inspire our admiration and fascination.

Light is the source of all beauty in gemstones. Interactions between minerals and light cause the intense colours of ruby and lapis lazuli, the sparkling fire of diamond and the play of rainbow colours in opal. Most gemstones show little beauty in the rough state and their full colour and lustre is revealed only by skilled cutting and polishing by a lapidary. Diamond's magnificent fire is displayed best in the most precisely cut and well-proportioned of stones.

When we wear jewellery our movements create a continuously changing relationship between the gemstones and the light falling on them, adding sparkle to their colour and fire. Spotlights enhance diamonds, rubies and emeralds, while soft lighting brings out the glow of amber and pearls.

ABOVE Modern labradorite carving, displaying rainbow iridesence, with inlay of precious opal by Kreg Scully (1.5 cm).

RIGHT Faceted sinhalite 15.5 carats.

Rarity

If beauty creates the initial impact of gemstones, then rarity imparts an aura of exclusiveness and worth that increases our desire to possess them. Rarity determines the fabulous values placed on the great gems of the world and is a major influence on the prices of the gems displayed in every jeweller's shop window.

Gems may be rare in one or more respects. Many gems are varieties of common materials, their rarity residing in an exceptional colour or clarity. Quartz and feldspars together make up about two-thirds of the Earth's crust, but most are grey or cream in colour. Very little quartz has the beautiful colour and flawless transparency of a fine amethyst, and only rarely does labradorite feldspar display rainbow iridescence (left). A few gem minerals are of rare occurrence. An example is diamond, which forms a minute proportion of the kimberlite host rock – about 5 grams in 100 tonnes. Other minerals contain rare chemical elements such as the

beryllium in taaffeite and emerald (see pp. 40–41). In a few exceptional gems these qualities are combined with uncommonly large size. The Cullinan diamond, for example, weighed 3106 carats in its rough state.

Gem values, like ideals of beauty, vary with fashions and beliefs in different cultures, fluctuating with public demand and the availability of a particular gem.

Durability

Gems endure through time because they are resistant to chemical alteration, are sufficiently hard to retain a good polish and are not easily chipped or broken. Ideally, they should be unaffected by the temperatures, pressures and abrasive dusts in our everyday lives.

Hardness is the measure of a gemstone's resistance to abrasion. The most easily and commonly used standard is Mohs' scale of hardness, which was set up in 1822 by the German mineralogist Friedrich Mohs (see p. 13). Mohs selected ten well-known minerals and numbered them in order of 'scratchability' from 1 to 10, so that a mineral will scratch all softer minerals that have a lower hardness number. Much dust and the grit in the environment (rather than just house dust, which is made up mainly of dead skin etc.) consists of quartz, which is number 7 on Mohs' scale, so a gem should have a hardness of 7 or more if it is to retain a good polish.

To wear well, a gemstone must also be tough. Despite being harder than quartz, emerald and zircon are brittle and chip easily, and diamond and topaz are among the many gems that may cleave (break) along planes of weak atomic bonding if dropped or knocked against hard objects. The toughest gem is nephrite jade, which has a hardness of 6 to 7. Nephrite's strength derives from its structure, which consists of a mass of minute, interlocking fibres. Because it is so tough, it can be made into the most exquisite bowls and intricate carvings (see p. 63).

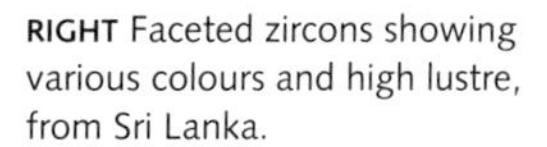
RIGHT Faceted zircons showing various colours and high lustre, from Sri Lanka.

Finding gems

The finding of a gem crystal or pebble can be an exciting event, especially when it leads to the discovery of a rich new gem source. Such discoveries are rare, for gem deposits occupy only a tiny proportion of the Earth's crust, their rarity stemming from the combination of physical and chemical conditions necessary for gem formation and transportation to the surface. The origins of gem minerals are as diverse as those of the rocks in which they are found.

Gems in the rocks

Many gem minerals crystallize at high temperatures and pressures deep in the crust and underlying mantle. Diamond forms more than 150 km underground, where temperatures are more than 1200 °C. Also from this region come some of the red pyrope garnets and rare green chrome diopsides.

Rocks at the crust/mantle boundary contain olivine in abundance, but only a fraction reaches the surface and only a tiny fraction of this is

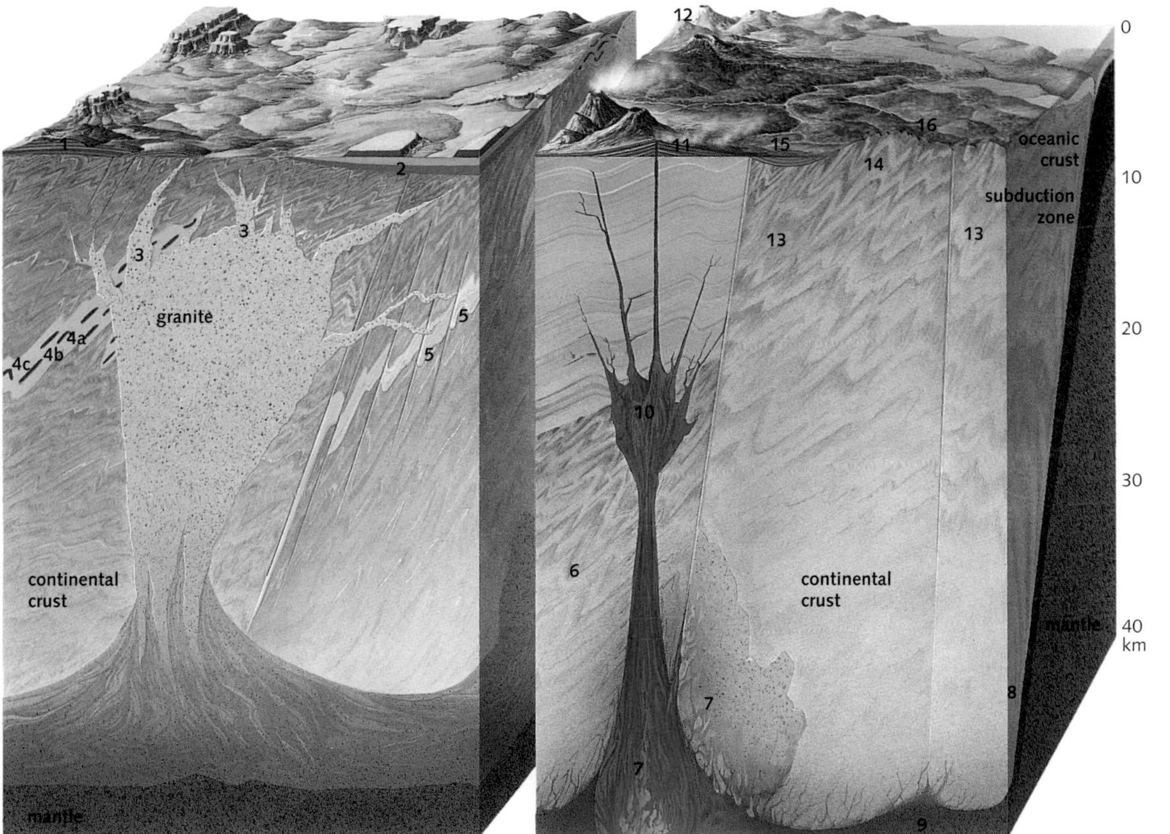

ORIGIN OF GEM MINERALS.

1 Some turquoise occurs as veins and nodules in decomposing volcanic rocks.

2 Opal in sediments forms by evaporation of silica-rich groundwater.

3 Aquamarine, tourmaline, topaz, quartz varieties, spessartine garnet, chrysoberyl, moonstone, scapolite and spodumene crystallize in pegmatites.

4 a) Ruby, sapphire, spinel and zircon, b) lapis lazuli, c) spessartine and grossular garnet, form in reactions between hot granitic fluids, impure shales and limestones.

5 Emerald forms by reaction of granitic fluids with chromium-bearing rocks.

6 Ruby, sapphire, spinel, chrysoberyl and garnets form by alteration of aluminium-rich, muddy sediments.

7 Peridot forms in basalts and ultrabasic rocks.

8 Jadeite forms at high pressures in subduction zones.

9 Pyrope garnet and diamond form in the mantle.

10 Gem minerals are caught up in rising basaltic magma.

11 Gem minerals reach the surface in basalt lavas.

12 Much amethyst, citrine, agate and opal is deposited from silica-rich fluids and fill gas cavities in lavas.

13 Gem-bearing rocks are uplifted to the surface.

14 Weathering releases gem minerals into the soil.

15 Gem minerals accumulate in river gravels.

16 Jadeite boulders accumulate in rivers.

gem-quality peridot (7). These rocks and minerals reach the surface either by uplift during mountain-building episodes (13), or in upwelling magmas (10). For example, rubies, spinels and jade formed at depth have been brought to the surface in mountain ranges such as the Himalayas, while zircons, garnets, rubies and sapphires are found in the basalt lavas (11) erupted from considerable depth, for example in eastern Australia or southern Thailand. Once at the surface, weathering processes (14) break down the rocks and release the gem minerals, which may then be washed downslope and concentrated by stream action in sands and gravels (15).

In the upper part of the crust pressures are low, but temperatures are more variable and depend on the proximity of basaltic or granitic magmas (4 and 5). Such magmas may be sufficiently hot to convert aluminous shales to slates containing garnet, iolite (cordierite) or andalusite. Fluids arising from magmas interact with limestones to form skarns (4), which are a rich source of rare minerals and gems such as lapis lazuli, grossular garnet and scapolite. During the final phases of crystallization of many granites, elements such as lithium, beryllium and boron, which are relatively rare in the crust, are concentrated into the residual fluids. With falling temperatures these fluids crystallize as pegmatite (3) and form the most important source of beryl, tourmaline, spodumene and topaz. Exquisite gem-quality crystals of these minerals have been found in the pegmatites of Brazil, California, Madagascar and the Ural Mountains. Pegmatites are distinguished by the large size of their crystals. Crystals of opaque beryl and spodumene may be several metres long, but gem-quality material is smaller, although still found in crystals weighing many kilograms.

Near the Earth's surface, magmatic fluids may interact with groundwater and cause considerable

RIGHT Amethyst crystals lining a cavity in rock.

movement of material by solution and precipitation. Such fluids may leach specific metals from orebodies and redeposit them elsewhere, forming secondary orebodies. The carbonates of copper (malachite and azurite) are formed in this way near copper orebodies. Turquoise (1) is also a secondary concentration of phosphate with copper and iron. Much quartz and chalcedony is deposited in fissures and cavities from silica-rich fluids. Most commercial amethyst and agate is found lining and filling cavities in lavas (geodes) (12).

In contrast to the gems formed from relatively rapidly moving fluids, precious opal takes shape as a collection of silica spheres under conditions of extreme stability and quiescence (2). Such conditions occur either in porous sedimentary rocks in places with a long history of crustal stability, as in Australia and Brazil, or occasionally in cavities in volcanic rocks, for example in Hungary, the Czech Republic, Mexico and Honduras.

Diamonds in the rocks

Diamonds are generally found either in pipe-like bodies of kimberlite or lamproite rock, or in river and beach gravels, where they accumulate after erosion of these pipes. All diamonds are valuable, either as gemstones or for use in industry, so they are mined in a much more systematic and mechanized way than other gem minerals. Kimberlite pipes range from a few metres to 1.5 km in diameter and may reach depths of 3 km. Only a small percentage of pipes contain diamonds and these are mined from open pits down to about 300 m, then by a variety of underground techniques, one of which is the block caving method. Mineable kimberlite contains on average about 25 carats (5 grams) of diamond per 100 tonnes of rock,

ABOVE Octahedral diamond crystal in kimberlite.

BELOW Jwaneng diamond mine, Botswana.

and of this 5 carats may be of gem quality. These minute quantities of diamond are separated from vast amounts of rock and sediment by exploiting the unique properties of diamond. Ores are first crushed and the heavy minerals, including diamond, are concentrated. The heavy minerals may be passed onto a grease belt where a stream of water washes away all material except the diamonds, which are 'non-wetable' and stick to the grease. Alternatively the heavy minerals are passed through an X-ray beam, which causes most diamonds to fluoresce. The fluorescence causes another instrument to emit a blast of air, which diverts the fluorescing diamonds into a separate container. The final check is done by hand.

Rough diamonds are first separated into gem and industrial qualities. Gem-quality crystals are then sorted according to their size, shape, the extent and position of flaws (inclusions, cracks or imperfections), and colour.

Well-travelled diamonds

Diamonds, by virtue of their hardness, may be washed out of the rock in which they were formed and carried vast distances by water to settle in riverbeds and oceans. During the journey those diamonds that have flaws or cracks are broken and crushed. At the end of the journey, although the yield may vary, generally the quality of the surviving stones is good.

One area where diamonds have been concentrated after weathering and transport is on the Namibian coast. The Namibian beach deposits are mined in huge open pits, behind protective walls. These are built from sand that has been removed to uncover the ancient, diamond-bearing, beach terraces. The yield here is only about 5 carats of diamond for every 150 tonnes of rock removed, but almost all this is of gem quality.

BELOW Coastal diamond mine, Namibia.

Off the coast, large mining ships equipped with vacuum equipment collect diamonds from the sea bed. More than half a million carats of diamonds were recovered off the coast of Namibia in 1999. New mining vessels were introduced in 2000 that use a crawler to mine horizontally on the sea bed.

Gem gravels

Gem gravels are among the most productive sources of fine gems; many of the world's most famous gems were mined in the Golconda diamond fields of India or the gravels of Myanmar (formerly Burma) and Sri Lanka.

Gem-rich gravels form because most gem minerals are harder and more chemically resistant to weathering processes than the rocks in which they form. Gem grains either accumulate in the weathered surface layer of the parent rock, or are washed downslope and deposited in river gravels some distance away from their source. Since gem minerals are generally heavier than many common minerals, concentrations build up wherever river flow decreases in speed and volume. Flawed fragments are more readily destroyed when gem minerals are transported over long distances and in turbulent conditions, so some gravels contain a high proportion of fine-quality material.

The patchy distribution of gem material in any deposit, whether gravel or parent rock, makes a prediction of the potential yield almost impossible. As many deposits are also small or in remote and difficult terrain, large-scale mechanized mining is rarely economic. Unconsolidated gravels are the most easily worked of all gem deposits, and some gem gravels have been hand-dug for centuries. At Pailin, Cambodia the gravel is washed and sorted on a simple shaking-table (centre), and the sapphires are handpicked from the smaller residues (right foreground).

Mining in the Australian sapphire fields is more systematic and mechanized. Boreholes are drilled first to determine the extent and yield of the gravels. After removal by large earth-moving equipment, the gravel is broken up and sorted into size fractions in a trommel (rotating drum). The smaller material, which contains the sapphires, is passed onto pulsating jigs, which concentrate the heavy minerals. This concentrate is finally sorted and graded by hand.

Gems around the world

Exploration of our planet has revealed gem deposits in almost all countries. A few areas are rich in both the variety and quantity of gem minerals produced, including Minas Gerais in Brazil and Mogok in Myanmar. Other localities yield a single gem species of superb quality, as typified by Colombian emeralds. Most deposits are small and rapidly exhausted, but a few, such as those of Myanmar, Sri Lanka and Afghanistan, have been in production for many centuries.

TOP Sapphire-bearing gravel, Australia. The sapphires are derived from basaltic rock.

ABOVE Ruby and sapphire fragments from gem gravels.

LEFT Gravel workings in the ruby and sapphire fields at Pailin, Cambodia.

Properties

How do you identify a gemstone, given the huge range of possibilities? Experience helps, for the more stones you see and handle, the better you are able to assess such properties as colour, lustre and density. But even the most educated guesses need confirmation, and many jewellers rely on some simple instruments to do this such as the hand lens (x10 magnification), refractometer and spectroscope.

Pick it up and look

When trying to find out what it is, the first thing to do, if at all possible, is to pick up the crystal, gemstone, ring or piece of jewellery and hold it in your hand. Hold it up to the light, turn it, and most importantly, look. With a well-formed crystal, the crystal shape is probably the most useful characteristic. You may get a fair idea of what the crystal might be from its general appearance, its colour, whether it is transparent, translucent or opaque, how heavy it feels and whether there are any noticeable features such as growth marks or whether the surface is worn or scratched, or there are internal features such as inclusions. Once a gem-quality pebble, crystal fragment, or crystal has been cut and fashioned as a gemstone, the shape of the crystal is lost and further clues have to be found. Again an initial appraisal of colour, lustre, cut and density will help to narrow the field. Obviously if the gemstone is red, it cannot be an emerald. But there are many red gemstones, so it may not be a ruby. The way that the gem has been cut and the quality of the cut, the sharpness of the facet edges and the amount of wear and tear on the stone give clues about its hardness.

BELOW The change in lustre can be seen where the garnet top joins the glass base, across the bright area which is a single facet (flat polished surface) in a garnet-topped doublet.

Lustre and hardness

The bright or dull look of the surface of a gemstone is referred to as its lustre. The term lustre describes the amount of light reflected from the surface of a gem or mineral. Diamond shows a very high lustre, which is described as adamantine. Most gems have a glassy lustre (called vitreous lustre), while less lustrous surfaces, such as those of turquoise, nephrite and amber, are variously described as waxy, greasy and resinous.

The higher the refractive index of the stone and the greater the surface polish, the higher the lustre. However, some relatively soft gemstones can show a very high lustre, whether or not they are polished. Generally the harder the gemstone,

MOHS' SCALE OF HARDNESS

The complete scale, in order of increasing hardness, is talc (1), gypsum (2), calcite (3), fluorite (4), apatite (5), orthoclase feldspar (6), quartz (7), topaz (8), corundum (9) and diamond (10). The intervals between the numbers do not represent equal increases in hardness. The difference in hardness between diamond and corundum is far greater than that between corundum and talc. A better comparison is given by indentation tests, such as the Knoop test, where dents produced by a diamond point at controlled pressures are measured. Complicated and expensive equipment is needed to carry out the Knoop test, whereas the Mohs' test can be done by scratching one mineral against another. Hardness tests are destructive and should not, of course, be carried out unless deemed absolutely necessary.

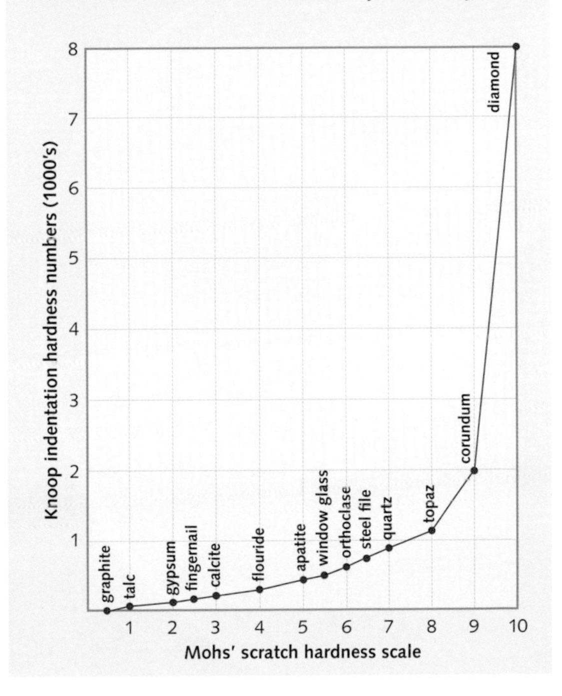

the better the polish that a gemstone will take. Harder stones will retain a good polish for longer than less hard gemstones. By looking at the condition and signs of wear and tear, such as scratched facets, rounded or chipped facet edges, and the general colour and lustre of the stone, it is possible to assess the general hardness and toughness of a gemstone without resorting to destructive scratch tests. A harder stone will have sharper facet edges, while a less hard or tough stone will generally have rounded or chipped facet edges.

How does it feel?

When handling gemstones regularly, a jeweller or gem dealer will be able to estimate the weight and density of the gems. Without this experience, most of us need to use instruments to measure these properties. The commercial value of a gemstone depends on the quality of colour, freedom from internal flaws and its weight. The weight of a gem is measured in carats (5 carats = 1 gram), and gems are usually sold by weight at so much per carat. When handling stones of the same size,

LEFT Badly worn zircon facets.

some feel heavier, because they have a greater density, or specific gravity (SG). The SG of sapphire is higher than that of quartz, so when comparing the weight of a sapphire and a quartz of the same size, the sapphire will be heavier. The SG is measured by comparing the weight of the gem with the weight of an equal volume of water. The smaller the stone, the more difficult it is to gauge the SG, and testing with specialized equipment will be needed to give an accurate figure. Measurement of density is a useful tool to identify carved or large specimens of opaque or translucent gem materials and can also be carried out on gemstones that have not been mounted in jewellery and on crystals or crystal fragments.

Look closer

Having assessed the condition of the gemstone with the eye, the next step would be to examine the gemstone with a hand lens (loupe). You may be able to see many significant internal and external features if the gemstone is first cleaned free of dust and grease. With x10 magnification the gemstone appears to be ten times as large as it really is. This magnification is enough to reveal the quality of the cut. A x10 magnification will also reveal surface markings or textures that will help assess the durability of the gem (how hard and how tough it is). In addition you will be able to see the 'orange peel' surface on jade carvings and the uneven distribution of dye on

RIGHT Colour zoning in sapphire, from Sri Lanka.

some treated stones. Look at the external surface first and assess the wear and tear by looking for scratches and chipped areas (see p. 13). The surface of a crystal or crystal fragment may have deep scratch-like markings (striations) that were formed as the crystal grew or may have deep rectangular or triangular pits called etch-pits. The orientation of the striations may help in identifying the gem material. In some gemstones striations run along the length of a crystal (vertical striations); in others they are horizontal.

Looking into the gemstone you may see inclusions, cracks and flaws. A study of the inclusions (see p. 28) may yield clues to the identity and origin of a stone. Zones of colour, which are a growth feature, may give clues as to the crystal shape and to its origin. Straight growth zones (see opposite page) are usually seen in natural gems, whereas bubbles and curved layers are typical of glass and some synthetics. In doubly refractive gems, the back facet edges may appear doubled when viewed through the stone (see below right).

The building blocks – crystal structure

The shape of beautifully formed crystals is an outward expression of their orderly internal atomic structure. Under ideal growth conditions, crystalline substances would be built of units identical in shape, size and chemical composition. The shape of the basic units (the building blocks) controls the symmetry and overall shape of all crystals. A mineral crystallizes in one of seven symmetry systems. In nature, growth conditions are seldom ideal, so structural defects and chemical impurities occur in almost all minerals. However, all crystalline minerals possess an orderly atomic structure, whether they form as crystals, as outwardly shapeless masses or are found as water-worn pebbles.

Crystal structure influences many mineral properties that are important in cutting and identifying gemstones. For example, atoms may be less strongly bonded in some crystal planes, giving rise to directions of easy breaking or cleavage. Hardness may also vary with crystal direction. Optical characteristics used in identifying gemstones are a result of the interaction between light and the crystal structure of the gemstones.

LEFT Glass bead coloured to imitate agate, contains bubbles.

BELOW LEFT Double refraction in sinhalite viewed through the front of the stone (table facet).

Passage of light

The colours that we see depend upon how light travels through a gemstone. This is determined by the crystal structure, and the metallic elements present in minute amounts in the gemstone.

Making colours

Many gems appear coloured because part of the white light travelling through them is absorbed within the mineral structure. White light is a mixture of many colours, but only when one or more of these colours is removed does the light emerging from a gem appear coloured. The causes of absorption are complex, generally involving the presence of particular chemical elements and damage or irregularities in the crystal structure.

Most gems are coloured by a limited range of metals, the most important of which are chromium, iron, manganese, titanium and copper. Chromium gives the intense red of ruby and brilliant greens of emerald and demantoid garnet, while iron causes the more subtle reds, blues, greens and yellows in almandine garnet, spinels, sapphires, peridots and chrysoberyls. The most prized blue sapphires are coloured by titanium with iron. Copper gives the blues and greens of turquoise and malachite. Manganese gives the pink of rhodonite and orange of spessartine garnet.

In most gems these metallic elements occur as impurities, usually in minute amounts. Such gems can show a wide range of different colours, and because they contain such small amounts of impurity the colour of some may be altered, enhanced or destroyed by heating or by irradiation with gamma rays and high-energy sub-atomic particles.

In a few gems the colouring elements form an essential part of the chemical composition, for example the copper in turquoise, manganese in rhodonite, and iron in peridot and almandine garnet. These gems have a very limited colour range, generally restricted to shades of one colour (see p. 51). Such colours are stable and impossible to alter greatly without destroying the mineral.

More than one colour

Crystal structure affects the way in which light travels through a substance. In all minerals, other than cubic and non-crystalline minerals, light entering the mineral is split into two rays that travel at different speeds and along different paths through the crystal structure.

In coloured minerals the rays may be differently absorbed within the crystal structure and emerge as two or three different colours or shades of the same colour. This effect is called pleochroism, and can be particularly helpful in identifying gemstones. Pleochroism causes the directional variations in colour seen in many gem minerals, so a gemstone looks a different colour or shade of colour when turned and looked at from different directions. Viewing these different colours is made easier with the use of a small instrument called a dichroscope, which enables two colours to be seen at the same time through the eyepiece whilst turning the gemstone. Dichroic gemstones have two colours and trichroic gemstones have three different colours or shades of colour when viewed from different directions.

Rainbow colours

The brilliant colours displayed by opal and labradorite, and in the fiery sparkle of diamonds,

Use of filters to show the pleochroic colours of tanzanite (*top*), andalusite (*left*) and tourmaline (*right*).

LEFT Seen without filter.

LEFT With polarizing filter.

LEFT With polarizing filter at 90° to that above.

arise when white light is split into its constituent colours, the colours of the rainbow. White light consists of electromagnetic waves of different wavelengths, each wavelength appearing a particular colour. A complete 'rainbow' spectrum exists, from the long red wavelengths through to the shorter violet wavelengths.

Fire (dispersion)

Dispersion is the origin of the 'fire' in gemstones. The 'fire' of a gemstone is seen as flashes of the colours of the rainbow as the gemstone or the light source are moved. When light enters a mineral the various wavelengths are differently refracted (bent), red the least and violet the most, so that the colour spectrum is spread out. Gem minerals vary greatly in their ability to disperse light, and their dispersion can be measured as the numerical difference between the refractive indices (see p. 19) of specific blue and red wavelengths.

Play of colour

Interference causes the iridescence in labradorite and the rainbow effects seen in cleavage cracks and on tarnished surfaces. When light falls on very thin mineral layers, as in labradorite, it is reflected from both the upper and lower surfaces. Since the reflected rays have travelled different distances, the wave troughs and peaks of the various wavelengths either coincide or are out of step. A colour is enhanced if they coincide but little or no colour is seen for out-of-step wavelengths.

In precious opal, which is composed of transparent, regularly sized and stacked spheres, light is scattered by the network of spaces between the spheres. Interference occurs between the emerging rays, the range of colours seen depending on the size of the spheres and the angle at which the opal is viewed. Larger spheres produce a complete spectrum as the opal is tilted, but small spheres generate only blues and violets.

Measuring refractive indices

Each gem mineral interacts with light in constant and measurable ways, and these properties can be used in its identification.

Light entering a mineral is slowed down and refracted from its original path in air. Cubic and non-crystalline minerals are singly refractive, meaning that light is slowed and refracted equally in all directions through the mineral. Minerals crystallizing in all other systems are doubly refractive. In doubly refractive minerals the light entering is split into two rays that are differently slowed and refracted, and which travel along different paths through the crystal structure. Double images may be seen through minerals such as calcite, where these differences are large.

A constant mathematical relationship, called the refractive index (RI), exists between the angle at which light strikes a mineral and the angle of refraction. Singly refractive minerals have only one refractive index. Doubly refractive minerals show a range of refractive indices between an upper and lower limit characteristic of each mineral. The numerical difference between the maximum and minimum indices is called the double refraction or birefringence. Birefringence can only occur in doubly refractive minerals and is given the abbreviation DR.

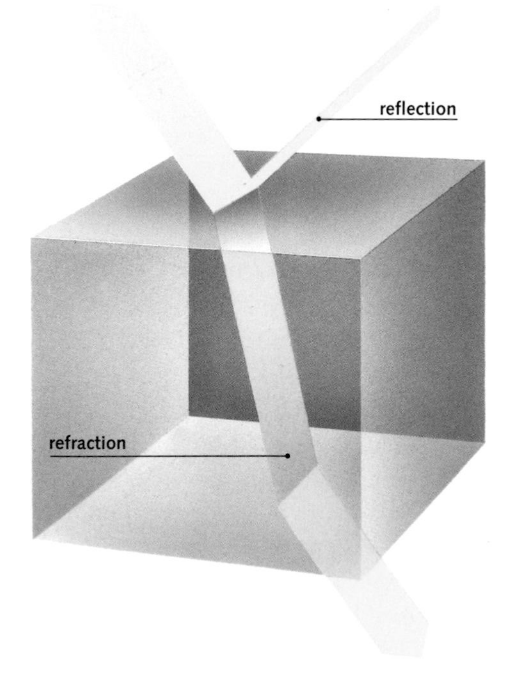

BELOW LEFT Reflection and refraction.

BELOW Double refraction in calcite.

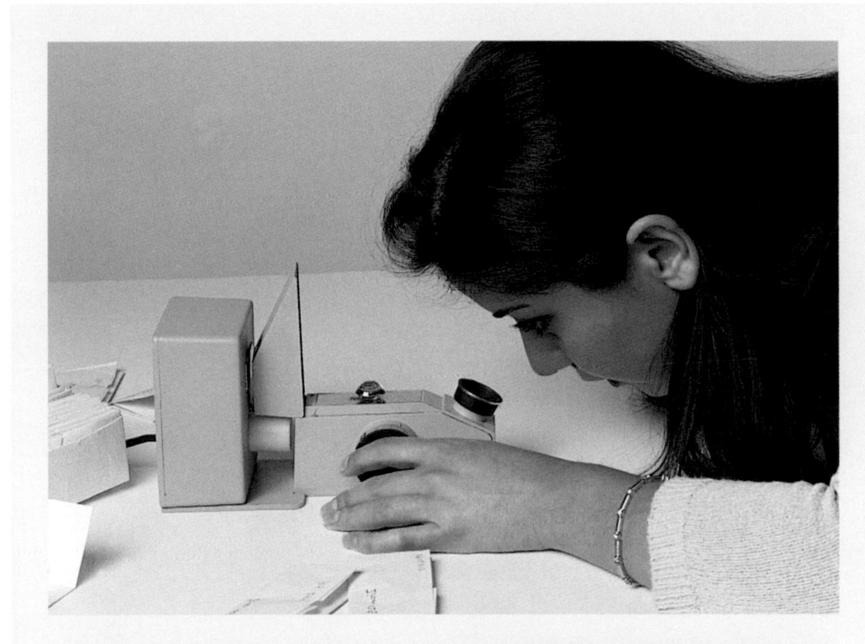

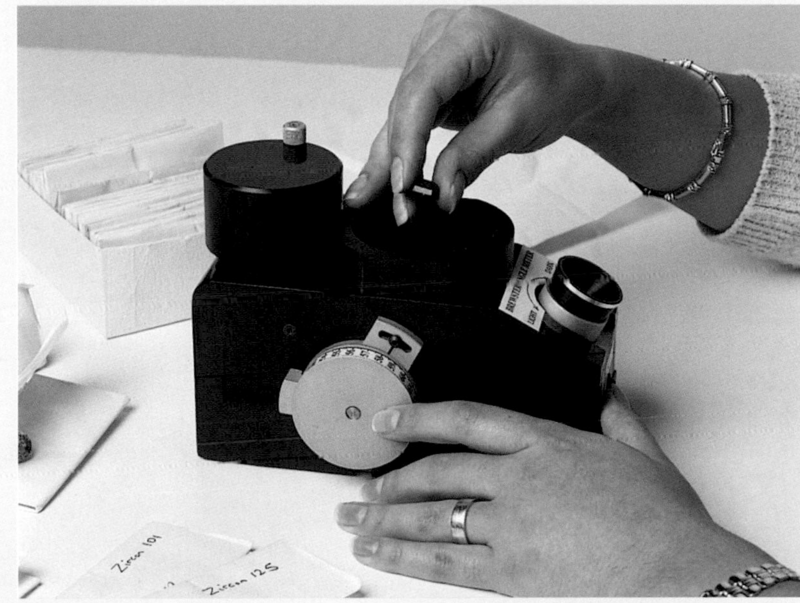

REFRACTOMETERS

The refractive index (RI) of most gemstones can be measured accurately on a standard gemmological refractometer if the gem has at least one flat, polished surface. This surface is placed on a glass prism in the top of the refractometer and light, usually of a single yellow wavelength, is shone into the instrument. If the gem has a lower RI than the prism glass, some of this light is refracted out through the gem and the remainder is reflected back into the refractometer. When you look into the eyepiece, the junction between reflected and refracted light appears as one or two shadow edges, depending on whether the gem is singly or doubly refractive. The position of the shadow edges depends on the refractivity of the gem, so they can be read off against a scale to give the RI of the gem, within the range of 1.4 to 1.8. Many gems have a unique combination of refractive index and birefringence (the difference between the maximum and minimum RIs).

A new instrument called the Brewster Angle Meter was introduced in 1999. With a laser as the light source, the instrument is not restricted to a range of 1.4 to 1.8 and can be used to find RIs of gems outside this range. This is particularly useful for diamond and diamond imitations such as CZ (cubic zirconia) and zircon that have RIs higher than 1.8.

SPECTROSCOPE

The Spectroscope is used to study the colouring element in gemstones (see p. 25).

ABOVE Refractometer in use.

LEFT Placing a gem inside a Brewster Angle Meter.

Cutting and polishing

A skilled lapidary can turn a rough pebble into a sparkling and valuable gemstone. The knowledge needed for this transformation has been built up over many centuries, and today a style of cutting can be selected to display the special qualities of each and every gem.

When deciding how best to cut a gemstone, the lapidary must consider the shape of the rough material and the position of flaws, fractures and inclusions. He or she must also be aware of the mineral's optical properties and physical properties such as cleavage. It is difficult to produce a good polish parallel to cleavage directions. Pleochroic gems (those that show a different colour or shade of colour from different directions) should be oriented to show their best colour. However, the cut is often a compromise between displaying the full beauty of a mineral and producing the biggest gemstone possible, since size also affects the value.

Cabochons are the oldest, simplest cuts. Cabochons are round or oval gems with plain, curved surfaces. Still in use today, cabochons display best the colours and patterns in opaque and translucent stones, and optical effects such as sheen, iridescence, cat's-eyes and stars.

The faceted styles now used for almost all transparent gems developed much later, becoming important in medieval Europe and India. In these styles the surface of the gem is worked into a pattern of highly polished, flat planes (facets), which act as mirrors. Some light is reflected from the surface of the crown (top) facets, displaying the lustre. Light entering the gem is reflected back through the top of the stone from the pavilion (bottom) facets, displaying the colour and fire. The facets must be precisely angled to bring out the maximum beauty, the angles varying according to the optical properties of each gem mineral. In badly cut stones light leaks out through the pavilion so that colour and fire are lost.

The brilliant and step cuts are the most familiar styles in modern jewellery. Brilliant cuts were developed to show off the superb lustre and fire of diamond, and are also used for many other gems. Step cuts are most effective in stones such as emeralds, where colour is the supreme quality.

FACETED CUTS

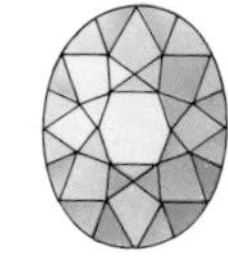

oval brilliant

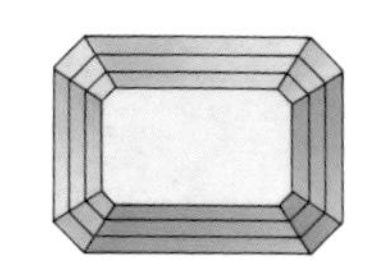

emerald or modified step cut

pear-shaped brilliant

marquise or navette cut

baguette

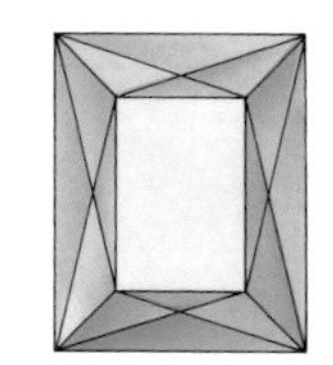

scissors or cross cut

ABOVE Styles of cutting.

TOP Six-rayed star sapphire.

LEFT De Beers' Millennium Star Diamond was on display in the Millennium Dome, London, UK, throughout 2000.

OLD CUTS

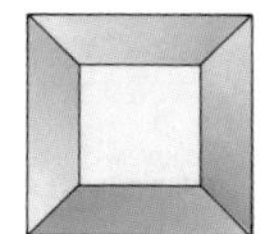

table cut

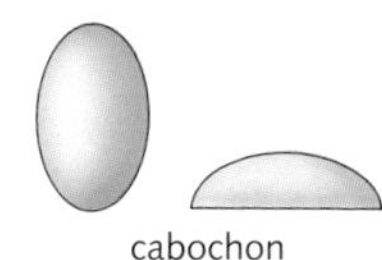

cabochon

round rose cut

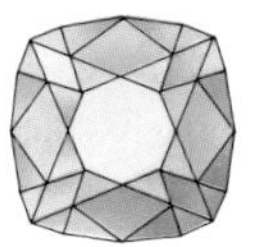

old mine cut

THE BRILLIANT CUT

The brilliant cut is probably the most important type of cut used for gemstones. The number of facets (flat polished faces) and the angles between them are worked out mathematically so that the gemstone looks bright and sparkles. Most diamonds are cut as round brilliants. Each has 58 facets (or sometimes more in a large stone), of which 33 are on the top part of the stone (the crown) and 25 are on the lower part of the stone (the pavilion). The point at the back of the stone may be removed to prevent accidental damage and to leave a small flat circular facet called the culet. Without the culet, the gemstone has only 57 facets.

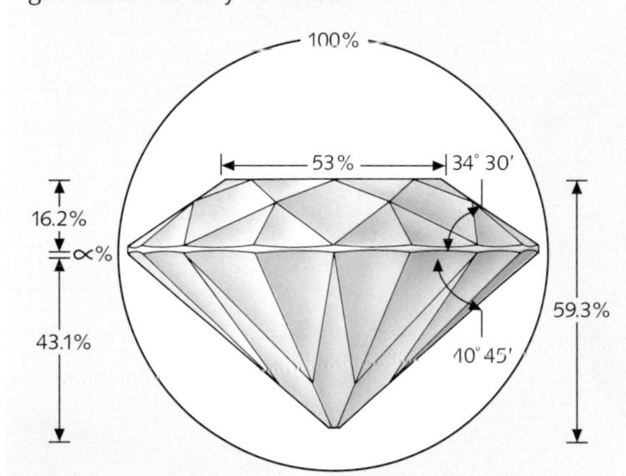

STAGES IN PRODUCING A BRILLIANT-CUT DIAMOND.

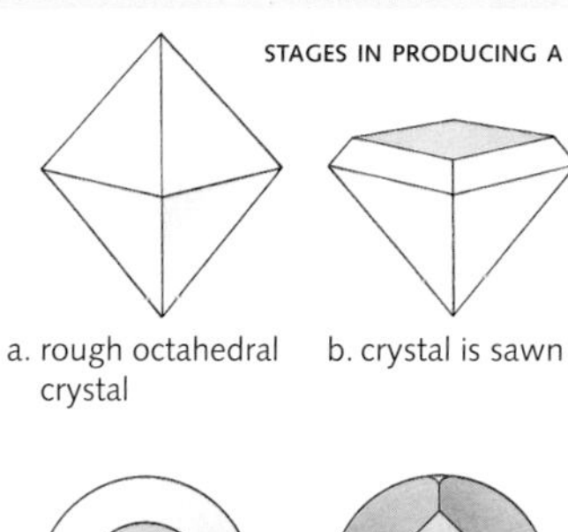

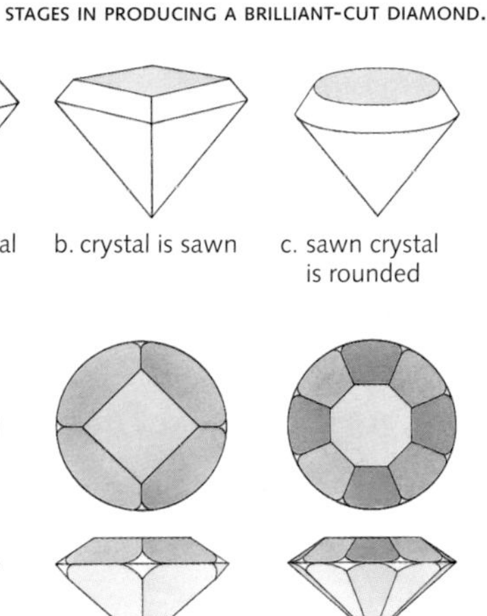

a. rough octahedral crystal

b. crystal is sawn

c. sawn crystal is rounded

d,e,f. Table (orange), bezel facets (pink) and pavilion main facets (blue) are ground and polished. The pavilion point is ground flat to form a culet.

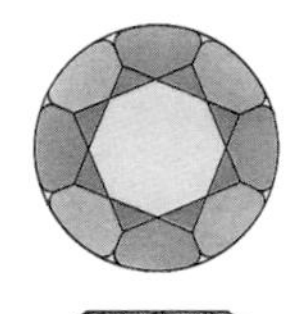

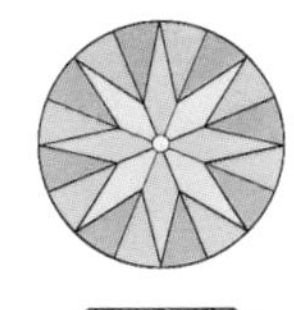

g. star facets are added to the crown

h. upper girdle facets complete the crown

i. lower girdle facets complete the pavilion

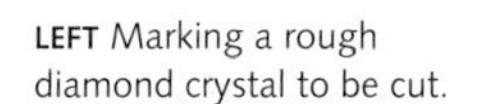

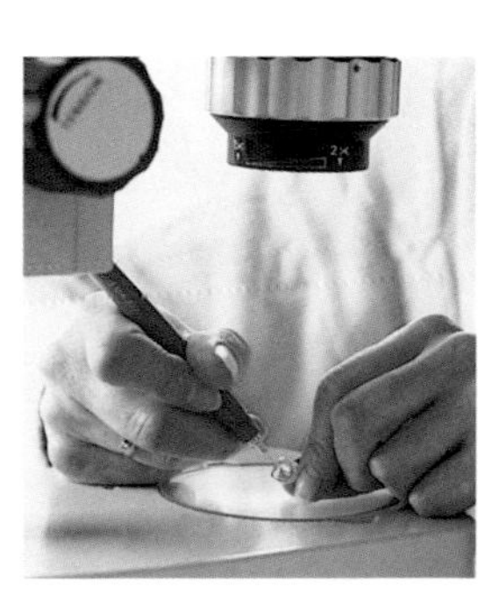

LEFT Marking a rough diamond crystal to be cut.

MIDDLE Sawing a rough diamond crystal using a spinning disc coated in diamond abrasive.

RIGHT Diamonds are faceted on a revolving scaife – a cast-iron wheel coated with diamond powder paste.

Real and fake

Any attractive object is rarely in the market-place for long before it is copied or imitated. Gemstones have been copied for at least 6000 years by a variety of materials, which can be described as imitations, composite stones or synthetics.

Imitation gemstones

Imitation gems look similar to natural stones, but are usually very different in their composition and optical and physical properties. They may be artificial substances or natural minerals of similar colour to the desired gem. Glass is a favourite simulant used to imitate many different gemstones, because it can be made in almost any colour and either moulded or cut to shape. Most glass, however, is much softer than the gems that it imitates and becomes badly chipped with wear. It may also contain bubbles (see below) and display a distinctive treacly or swirly texture. Glass is singly refractive, with a refractive index ranging between 1.5 and 1.7 – no singly refractive gem minerals fall within this range.

The most widely used diamond simulant of the last 20 years has been cubic zirconia (CZ), which was originally produced for use in laser and electronics research; more recently synthetic moissanite has come on the market. With their high refractive indices and fire, both are difficult to detect by visual tests alone. CZ can be distinguished from diamond by its lower reflectivity and heat conductance. Heat conductance is measured using a thermal probe.

Introduced in the late 1990s, synthetic moissanite, a silicon carbide (SiC), is a new synthetic gemstone with properties that are closer to diamond than any other imitation. With a

BELOW Imitation opal (Slocum stone).

BELOW RIGHT Faceted glass with bubbles.

hardness of over 9 on Mohs' scale, it is harder than ruby, sapphire or any other gem, except diamond. However, there is one main visible difference in that diamond is singly refractive and moissanite is doubly refractive. Because of this, a doubling of the pavilion facets can be seen on larger stones. However, this is difficult to recognize without the use of instruments when the stones are particularly small and set in jewellery. The RIs of moissanite are 2.648 and 2.691, dispersion is 0.104, birefringence is 0.043 and specific gravity is 3.22 (see p. 37).

Treated and enhanced stones

A number of techniques are used to improve the colour and/or appearance of natural and synthetic gemstones. The purpose is to increase their beauty, desirability and saleability. Probably the oldest method is that of heat-treating a gemstone to improve or change its colour. The heating of carnelian has been carried out in India for over 4000 years and oiling of emerald has been known for over 2000 years.

As a result of recent advances in technology, there are now many different techniques, which use modern equipment such as lasers, and computer-controlled heating and irradiating procedures. Lasers are used to drill holes into diamonds to reach inclusions. These are then evaporated or removed using chemicals, before the crack is filled. Some treatments are permanent, such as drilling, while others may be relatively temporary; for example, stains and fillings may leak, and some heated and irradiated stones may fade or revert to their original colour.

Most rubies and sapphires are heat-treated to improve their colour. Sapphires considered too dark can have their colour lightened by heating to 800–1400 °C in oxidizing conditions (with oxygen present). The very pale brownish-grey material from Sri Lanka that is called geuda can be changed to a blue by heating to temperatures of 1500–1900 °C in reducing conditions (without oxygen present). Variations in temperature and conditions allow more subtle colour changes, some of which only reach just

LEFT Imitation coral.

MIDDLE Imitation turquoise.

RIGHT Imitation lapis lazuli.

beneath the surface, while others alter the whole stone. For over 100 years brown topaz has been heated to give a more attractive pink, and amethyst (the purple variety of quartz) has been altered to the less common citrine (orange-brown variety).

As well as heating, gemstones can be irradiated to improve or change their colour. They may be exposed to gamma rays or bombarded by particles such as electrons, neutrons, protons or alpha particles. Much colourless topaz is irradiated and heat-treated to blue (see p. 49).

Most emeralds have flaws or cracks that detract from their beauty. The traditional method of oiling emeralds is a simple process. Essentially, it just involves immersing a stone in oil or wiping the surface with an oily cloth. The oil is then drawn into the cracks, with the result that they are less noticeable and the stone appears to be clearer and of a better colour.

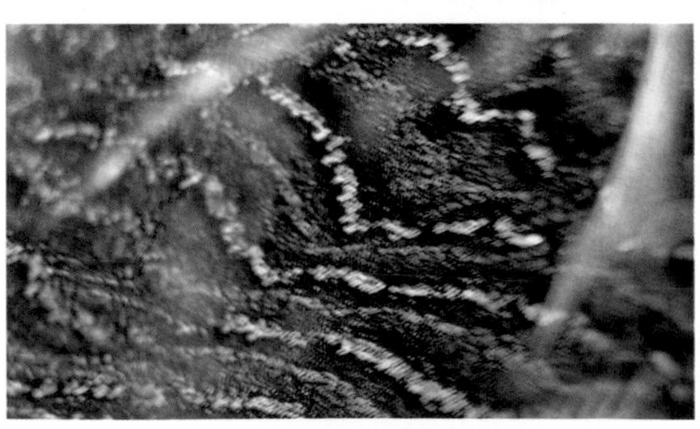

ABOVE LEFT Thai ruby (x25) showing typical 'fried egg' inclusion; a glassy spherical void with lace-like fringe.

LEFT 'Tiger's stripe' structure in amethyst.

Nowadays various colourless oils, waxes and plastics are used on a number of different gemstones. Some remain liquid; others, such as resin, set hard within the stone or as a surface coating. Turquoise, lapis lazuli, jade and some chalcedonies are dipped in liquid paraffin wax or given a surface coating of wax, after polishing, which penetrates the stone to fill cracks and gives a better surface colour. In addition, coloured oils and resins are also used. Matching the colour of the oils or resins to the stone improves the colour as well as hiding the cracks.

Where a stone has been oiled it may feel 'oily' or may leave a stain when wiped with an absorbent material such as a tissue. Years of wear or cleaning with ultrasound may displace any oils and fillings, with the result that the cracks in the stone will become more obvious and, in the worst case, the stone will fracture.

Coloured dyes and stains can also be used on some gems. Agate is dyed to imitate many gems or to give bright, but rather unnatural-looking pinks, greens and blues for decorative carved pieces. Quartz rocks have been dyed green to imitate jade and red to imitate ruby.

Foiling of stones involves placing a piece of reflecting material, such as a metal foil, behind the stone to change or improve the colour and make the stone appear brighter. Foiling was used in Britain, particularly during the Victorian era, to enhance costume jewellery made of paste (glass). Thin films of gold, silver and other metals can be deposited on the surface of gemstones and crystals to give a surface 'bloom'. When the back of the stone is coated, the mirror-like qualities increase the reflectivity and the stone appears brighter as well as taking on the colour of the coating. Quartz crystals coated with a surface film of gold to give a pale blue colour are called 'aqua aura'.

Composites

Composite stones (made of more than one part), have been made since Roman times. Doublets are the most common composites, consisting of two parts cemented together. Opal doublets are usually openly described and sold as such, but doublets consisting of a coloured glass base and hard mineral top are made to deceive. Garnet-topped doublets are made to imitate gems of all colours, but the junction between garnet and glass is often easy to see, emphasized by the differences in lustre or bubbles at the garnet/glass junction.

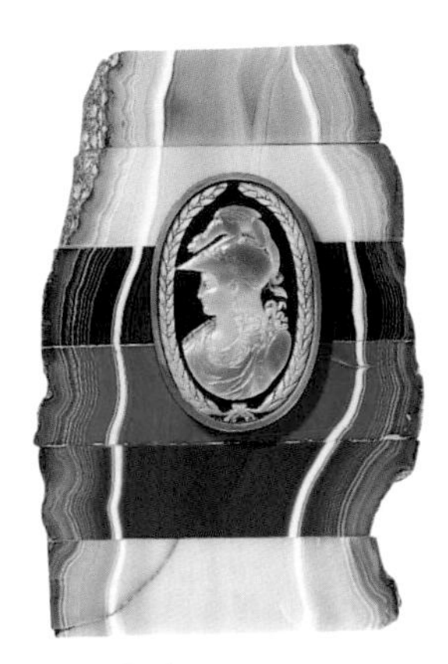

ABOVE Sardonyx cameo resting on dyed agate slabs.

BELOW Opal doublets.

ABOVE Spectroscope in use.

SPECTROSCOPE

The small spectroscope is useful in distinguishing some gems and imitations of similar colours. Light entering the instrument through a narrow slit is separated into its spectrum of colours (the colours of the rainbow) by glass prisms or a diffraction grating. If a gem is placed between a strong light source and the slit, dark bands may appear in the spectrum where these light wavelengths are absorbed within the stone. The wavelengths absorbed vary according to the colouring element (for example, iron, chromium etc.), so that the absorption spectra of gems of apparently similar colours may be very different. This test can be used on cut stones and rough material.

Soudé emeralds consist of a quartz crown and pavilion, between which is a layer of emerald-green glass or gelatine. The composite nature of these stones is seen clearly if they are immersed in water, but detection may prove difficult when the stones are set in jewellery.

Synthetics

Synthetic gemstones are almost exact copies of natural gem minerals. Made under laboratory conditions, most are manufactured by melting or dissolving the appropriate mineral ingredients and colouring agents, then allowing the molten mass or solution to crystallize at strictly controlled pressures and temperatures. The resulting crystals are virtually identical in both composition and crystal structure to the natural gem mineral, so possess similar optical and physical properties.

The earliest gem-quality synthetics were the rubies produced in 1902 by the Frenchman Auguste Verneuil, using a flame-fusion process.

ABOVE Faceted synthetic emerald.

BELOW LEFT Garnet-topped doublet.

BELOW Crystal of synthetic emerald.

BELOW RIGHT Boules of synthetic ruby and synthetic sapphire.

Synthetic spinels and sapphires followed soon after, and this method has proved so cheap and fast that it is still used to produce most synthetic rubies, sapphires and spinels. Emeralds, however, are made by other processes, and may take nine months to crystallize from a melt. Because of this, synthetic emeralds are more expensive, but may still be ten times cheaper than good natural stones. Today, as technology develops, it is possible to synthesize more and more gemstones, including opal, chrysoberyl and diamond. Synthetic gemstones can also be used to imitate other gemstones; for instance, aquamarine has long been imitated by blue synthetic spinel.

Because of the differences in their values, it is essential to be able to distinguish between natural and synthetic gems. Fortunately, some production methods may give rise to distinctive growth structures and inclusions, such as the curved growth zones and bubbles visible in many Verneuil-type synthetics, or the twisted 'veils' of

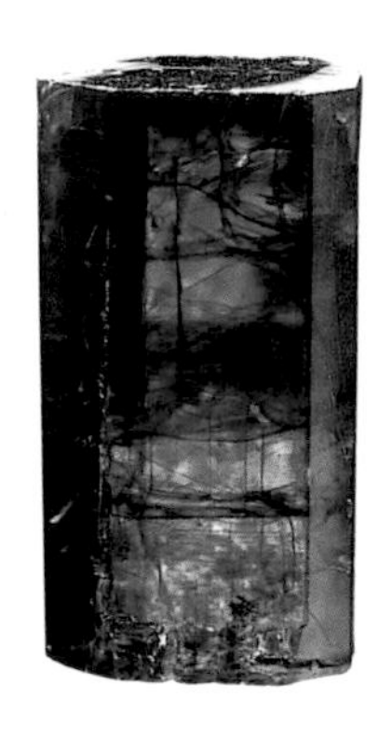

fluid-filled tubes in some synthetic emeralds. Many synthetic opals display a fine scaly pattern, the 'lizard skin' effect. Flawless gems, however, pose far greater problems, so the gemmologist must resort to complex instruments and techniques, such as infrared spectroscopy, for the answers.

Inclusions

Fine colour combined with flawless transparency is the ideal of beauty in many gem species, but as gemstones form they may trap other minerals. These mineral inclusions sometimes detract from the beauty of the stone, but they may also form the chief attraction in some gems. Inclusions cause the colour spangling in aventurine quartz and sunstone feldspar, and the cat's-eyes and stars that gleam from some chrysoberyls and sapphires.

To many people inclusions are merely flaws that reduce the value of a gemstone, but to the mineralogist and gemmologist they can reveal the gem's identity, how it formed, and even the source locality. Since the appearance of synthetic gems, inclusions have acquired a greater importance, often providing proof of a natural or artificial origin. Most inclusions can be seen with a hand lens (x10 magnification), but are best studied under the more powerful magnifications of a microscope (x60 magnification), which opens up a whole new world of fascinating images of great beauty.

Many gems contain small crystal inclusions, usually of different mineral species from the host gem. These can give valuable clues to the temperatures, pressures and rock types in which they formed. For example, emeralds from many localities contain mica flakes derived from the mica schists in which they formed, and Colombian emeralds may contain distinctive three-phase inclusions. These consist of jagged cavities containing saline (salty) liquid, a salt

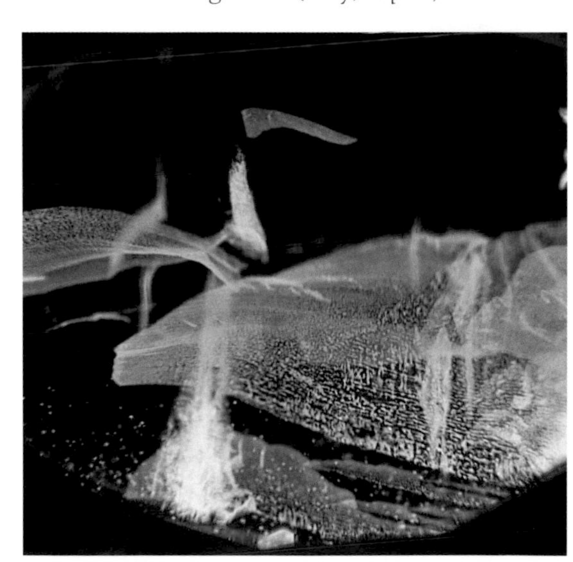

ABOVE Red garnet crystal in a faceted diamond.

LEFT Growth zones and bubbles in Verneuil synthetic ruby (x25).

RIGHT Wispy veils in Gilson synthetic emerald.

crystal and a gas bubble. Inclusions in diamonds may provide information both about their origin and about the mantle rocks in which they formed. The ages obtained from certain mineral inclusions indicate that some diamonds formed over 3000 million years ago and the youngest probably formed about 900 million years ago.

Crystal inclusions may have a good crystal shape, or they may be rounded. The distinctive internal granular texture of much hessonite garnet is formed by innumerable rounded apatite and calcite crystals. Slender hollow tubes and needle-like crystals of rutile, hornblende and asbestos occur in many gems, often developed parallel to one or more crystal directions; when abundant, they give rise to cat's-eye and star effects (see p. 20, top).

Fractures and cleavages may develop in minerals as a response to stresses during their crystallization or in later Earth movements. The 'lily-pads' seen in many peridots are stress fractures that develop around chromite crystals or other crystal inclusions. Moonstones sometimes contain insect-like structures which, on closer examination, are seen to consist of small cleavage cracks and are quite distinct from the real insects entombed in amber (see p. 71). Fractures may become partially healed during subsequent episodes of heating and alteration of the host rock, and many such fractures retain pockets of trapped fluid, as in sapphire 'feathers'.

LEFT Three-phase inclusions in Colombian emerald; each has a solid crystal and a gas bubble within a liquid inclusion in the gemstone.

BELOW RIGHT Rounded crystals and swirls in hessonite garnet give a 'treacly' appearance.

BELOW FAR RIGHT Rutile needles and 'feathers' in sapphire.

BOTTOM RIGHT A 'lily pad' inclusion typical of peridot; this specimen is from San Carlos, Arizona, USA.

BOTTOM FAR RIGHT 'Pseudo insect' inclusions in moonstone.

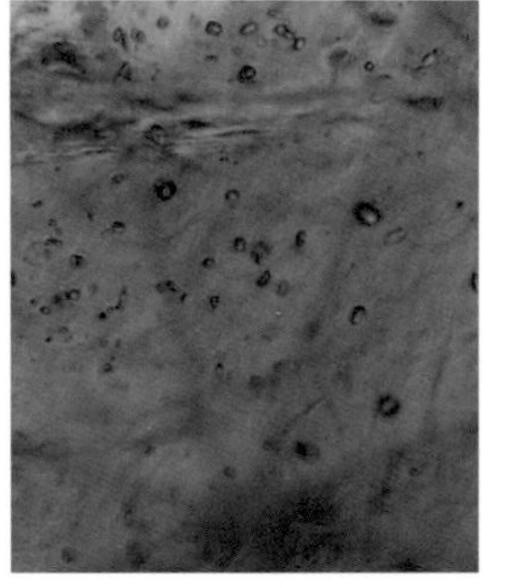

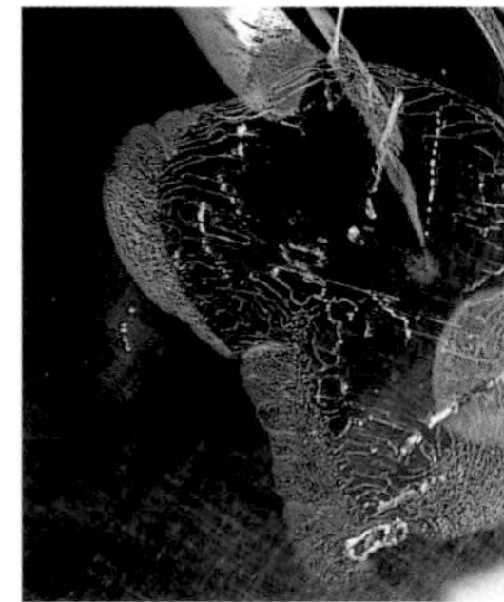

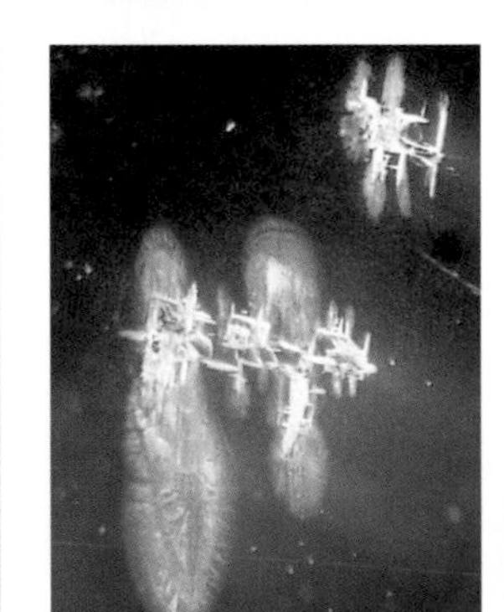

Jewellery

Natural objects such as seeds, berries, shells, bone and teeth used as adornment have been found in prehistoric sites. Fossilized shells used as beads on a necklace have been excavated from a site in Moravia dated at around 3000 BC. We do not know what significance these objects held for their owners, but in later ages gemstones have been used as religious symbols, to display wealth and status, and as good luck charms and amulets.

Ancient jewellery

The beauty and desirability of gold and precious stones inspired creativity and the development of the decorative arts. More than 4500 years ago, jade carving was already long established in China, and Sumerian and Egyptian craftsmen were making jewellery set with lapis lazuli, carnelian, turquoise, amethyst and garnet (see below left). Agate was particularly popular with the Romans, who skilfully used the differently coloured layers to make beautiful cameos (see below).

Many gemstones that were used in this early jewellery were brought great distances by travellers along trade routes. Beautifully coloured pebbles in stream-beds and on beaches were probably the first gems to attract the human eye and imagination. As civilizations evolved, a greater variety of gem minerals and more reliable sources of supply became available through organized mining and trading.

The Egyptians mined turquoise in Sinai and amethyst near Aswan, but they imported lapis

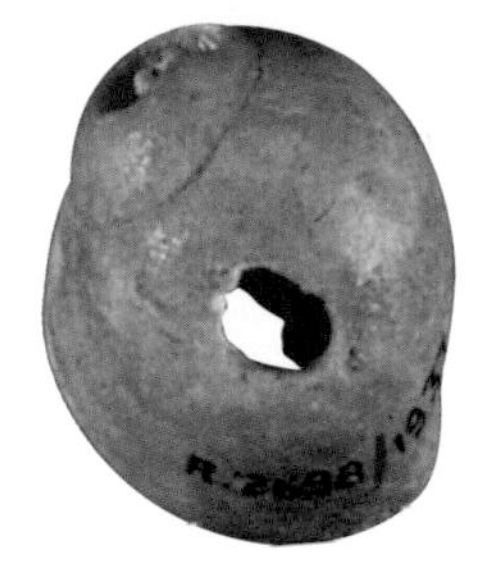

ABOVE Shells used as beads.

LEFT Sumerian court jewellery from Ur, set with lapis lazuli and carnelian.

RIGHT Cameo of the Emperor Augustus, first century AD.

lazuli from Badakhshan in Afghanistan, the only source known to the ancient world. These remote mines are perhaps the world's longest operating mines, still yielding the finest quality lapis lazuli after more than 6000 years. The Romans mined agate near Idar-Oberstein in Germany and, after centuries of neglect, these deposits again formed the basis of an important local industry from medieval times. Idar-Oberstein still has a thriving gem and jewellery trade, working on gemstones imported from all over the world.

The gem gravels of India, Sri Lanka and Myanmar are legendary for the rich variety of their

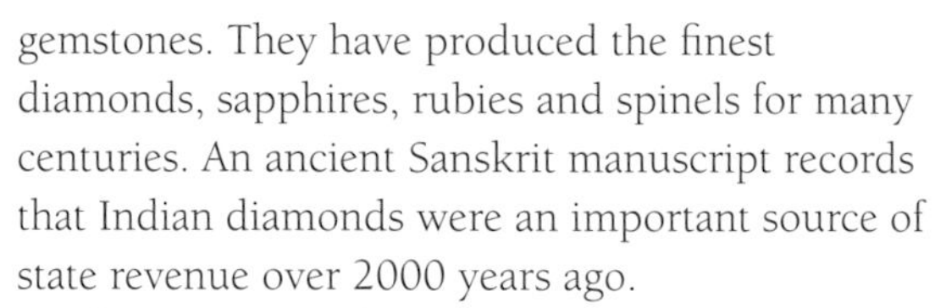

gemstones. They have produced the finest diamonds, sapphires, rubies and spinels for many centuries. An ancient Sanskrit manuscript records that Indian diamonds were an important source of state revenue over 2000 years ago.

The great gems from these sources have always exerted a powerful fascination. Some have retained an individual identity through many colourful adventures, acquiring a charisma beyond ordinary commercial values. When the Koh-i-Noor diamond was presented to the Mughal Emperor Babur in 1526, its value was set at "one day of the whole world's expenditure". A few of these gems even

LEFT Koh-i-Noor set in a Maltese Cross on the crown of Queen Elizabeth the Queen Mother.

RIGHT Spanish Renaissance pendant set with Colombian emeralds.

carry written evidence of an illustrious past, as on the Shah Diamond, which is inscribed with the names of three royal owners including Shah Jahan.

Antique jewellery

Jewellery more than 100 years old is generally referred to as antique, but the term 'antique jewellery' covers many different types, ages and styles. In Europe, the jewellery of the Middle Ages (medieval), which was often functional, included jewelled ring brooches and girdles, in addition to the religious icons and tokens of love. Renaissance jewellery (15th to early 17th centuries) was developed in Italy under the patronage of the noble families in the princedoms.

Exquisite gems have come from more recently discovered deposits in the Americas, Africa, Australia and Siberia. Magnificent Colombian emeralds first reached Europe in the 16th century, brought to Europe by the Spanish Conquistadors; the emeralds were of a better colour and size than those that had been previously mined in Habachtal (Austria) and Egypt, and perhaps from ancient Asian sources. Exploration of Brazil revealed gem deposits especially rich in topaz, tourmaline, chrysoberyl and agates.

LEFT Pair of Victorian diamond braclets (top) circa 1870. Two pairs of diamond pendent earrings (bottom), one on either side of a Victorian diamond brooch circa 1860.

ABOVE French Art Nouveau brooch of a gold lady with butterfly wings set with gems and enamel, by Gaston Lafite. The butterfly wings have diamonds and rubies (circa 1900).

RIGHT Buckle pin of gold and silver, with diamonds, pearls and ebonite, with the contrasting colours, simple lines and flat angular shape of Art Deco (circa 1925).

South African diamonds and Australian opals are two important 19th-century discoveries.

Another term in common use is 'period jewellery' which is recognizable, by its style or design, as coming from a particular period or time. Notable British period jewellery includes that designed during the reign of a particular monarch or dynasty, such as Georgian and Victorian. Other jewellery is associated with a particular group or movement, such as Arts and Crafts (1890–1914), or a distinctive era, such as Art Nouveau (1885–1915), Art Deco (mid 1920s–1930s) and Retro (1940s–1950s). Although the dates give an idea of the initial period of design and manufacture, modern jewellers continue to design and make jewellery that would be described as being, for instance, in an Art Deco style.

The Art Nouveau style was flamboyant and free flowing, in contrast to the earlier styles of mass production, and covered all areas of design. Many ideas were taken from nature, with floral patterns, butterflies and birds, and beautiful women in flowing gowns. Jewellery made use of cabochons of opal and moonstone, pearls and colourful enamelling, in addition to faceted stones.

Art Deco originated in France as an alternative to the flowing designs of Art Nouveau. The strong colours, mainly black and white, and the rigid geometric designs of Art Deco are in stark contrast to those of Art Nouveau.

Modern jewellery

In Europe, the gem and jewellery trade suffered greatly during World War II. In the austere post-war years some designers experimented with cheap materials. This led the way to a more experimental environment in the 1960s.

Jewellery was no longer constrained by the tradition of using metals and gemstones. Paper, wood and other material were used to supplement ideas, and gemstones could be used alone, without the surrounding mounts of metal. New technology and new materials led to an explosion in design.

Throughout the 1960s and 1970s and beyond, designers continued to experiment with designs and materials, looking for ways to shock, to amuse or to flatter the wearer or the audience. Some pieces are so extravagant that they could not be worn comfortably, while others almost melded to the skin and could be worn as everyday wear at home or in the office. A range of jewellery, from mass-produced rings and chains to expensive, individually commissioned pieces, can now be designed at a range of prices.

The 20th century has seen the expansion of the diamond industry into Siberia, Australia, Canada and many African countries. Newly discovered minerals and mineral varieties, such as charoite and tanzanite, have increased the range of materials available to jewellers, but the functions of our jewellery remain the same as those so important to our ancestors – to beautify and to impress.

LEFT Bracelet by Mouawad, set with rubies and diamonds.

BELOW Bracelet with amethyst, tourmaline, garnet and iolite, by Barbara Bertagnolli.

The gemstones

Most gemstones are minerals that have formed in a variety of environments within the Earth. Minerals have a definite chemical composition and ordered atomic arrangement, and therefore their physical and optical properties are constant or vary only within narrow limits. These properties can be measured accurately and are used to identify a mineral. In this part of the book, the main properties for each gem are listed as a boxed item. These lists can be used to compare one gem with another and to help with identification.

Gemstones can be grouped in a number of ways, for instance by colour and appearance, chemistry, value (which may change along with fashion) and use. The order adopted in this book takes all these aspects on board. The first gemstone described is diamond, probably the best known of all gemstones. It is the hardest and is unique among gems in being composed of a single chemical element – carbon. Other well-known gemstones such as ruby, sapphire and emerald are followed by the lesser known gems and those that are seldom cut for anyone but the collector of the rare and unusual. The gemstones covered up to this point in the book (p. 59) are all transparent and usually cut and polished as faceted stones, with many flat faces. A few show optical effects such as chatoyancy (cat's-eyes) and asterism (star stones), but only when cut as a cabochon.

The next section introduces the gemstones and rocks (made of one or more minerals) that are generally translucent or opaque (pp. 60–70). Little or no light can be seen through them, and they are usually cut as cabochons, fashioned as beads, cut as slices, used in decorative pieces such as inlay, or carved. They have attractive textures or patterns (for example malachite) or show optical effects such as chatoyancy or asterism.

Organic gems, those of an animal or plant origin, form a separate group at the end of the book (pp. 70–72). Plants and animals are the sources of the more fragile 'organic' gems that have been used for ornament since very ancient times. Jet and amber are the fossilized wood and resin of extinct trees; pearls, shells and most corals are calcium carbonate structures formed by animals that live in water. Ivories are the tusks and teeth of land and sea mammals.

LEFT Cross set with sapphires.

PROPERTIES

CC	chemical composition
CS	crystal system
H	Mohs' scale
SG	specific gravity
RI	refractive index
DR	birefringence

Diamond

Diamond is named from the Greek 'adamas', meaning unconquerable, an early recognition that it is the hardest of all natural minerals. This supreme hardness is combined with exceptional lustre and dispersion, giving diamond the lasting fiery brilliance for which it is prized.

It is perhaps difficult to believe that diamond, like graphite and charcoal, is a form of carbon. Diamond crystallizes in the cubic system, at enormous pressures and high temperatures. Its exceptional properties arise from the crystal structure, in which the bonding between the carbon atoms is immensely strong and uniform. Much diamond occurs as well-formed crystals (see below), most commonly as octahedra. Graphite, which has a hardness of 1 to 2 on Mohs' scale, consists of weakly bonded sheets of carbon atoms. Charcoal is non-crystalline.

Diamond is the most intensively mined and carefully graded of all gem minerals. The quality of a gem diamond is assessed by a system known as the 'Four Cs': colour, clarity, cut and carat weight. Diamond varies from colourless, through a range of yellows and browns, to green, blue, pink and a very rare red. Colourless diamonds or fancy coloured diamonds (those of strong or unusual colour) are considered the most valuable. Truly colourless stones are rare as most diamonds are

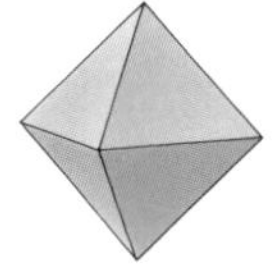

octahedron

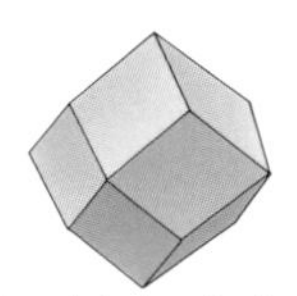

rhombdodecahedron

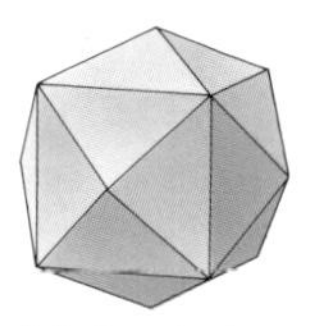

tetrahexahedron

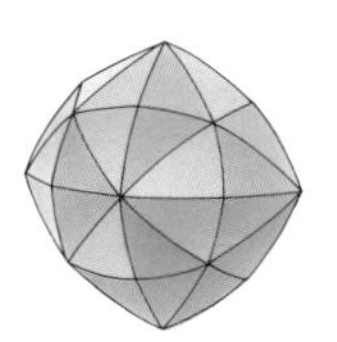

hexoctahedron with curved faces

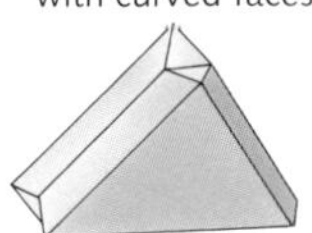

twinned macle

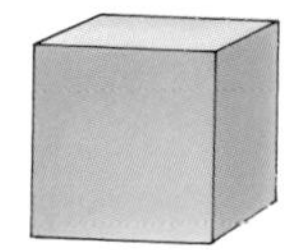

simple cube

ABOVE Some crystal forms of diamond.

LEFT Round brilliant-cut diamond surrounded by crystals of diamond.

ABOVE Pink diamond.

RIGHT Octahedral diamond crystal.

BELOW Display of faceted diamonds to show the range of colours.

tinged with yellow or brown, by impurities. The most common impurity is nitrogen, which gives rise to brown, yellow, green and black diamonds, depending on the amount present and its distribution in the crystal structure. Blue diamonds contain minute traces of boron.

Clarity is judged on the extent of mineral inclusions and flaws, such as cleavages, visible at ×10 magnification. Inclusions can affect the commercial value of gemstones either by reducing clarity or by providing valuable information about the origin of a gem. Studies of some garnet and pyroxene inclusions, coupled with a knowledge of the conditions needed to synthesize diamond, reveal that many diamonds formed at depths greater than 150 km underground.

The cut is of supreme importance in displaying the full beauty of diamond. Although diamonds were known in India 2300 years ago, crystals were not cut for many centuries because it was believed that diamond had magical properties and that these would be lost if the diamond was cut. In Europe, the polishing of simple point and table cuts from octahedral crystals, and rose cuts from cleavage fragments, began sometime after 1300. The popularity of diamond has grown with the development of the brilliant cuts, which best display the fire in diamond. Although early versions appeared in the 17th century, the modern form of this cut was set in 1919, when Marcel Tolkowsky published 'ideal dimensions' for a brilliant-cut diamond (see p. 21).

Today lasers are used to cut diamond, but the only mineral that can cut a diamond is diamond, and grinding and polishing are carried out by a lapidary using powdered diamond. This is only possible because diamond is less hard in certain crystal directions. Diamond also cleaves relatively easily, in four directions parallel to the octahedral crystal faces, so cleaving is sometimes used to divide large crystals and to trim away flawed material.

For over 2000 years, diamonds were found only as eroded crystals in river gravels. Until 1725 India was the major source of diamonds, with much smaller amounts mined in Kalimantan (Borneo). Diamonds were then discovered in Brazil, which became the leading supplier as Indian production waned. South African diamonds were found first in 1867, in gravels near the Orange River.

Further exploration in the Kimberley region of South Africa revealed volcanic pipes filled with a hitherto unknown rock type which contained diamonds. The rock, a variety of peridotite, was named kimberlite and was recognized as the diamond source rock. This discovery formed the basis of the huge modern diamond industry. Many similar pipes have since been found in other African countries, in Siberia, China and more recently in the Northwest Territories of Canada. A closely related rock type, lamproite, is the source of Western Australian diamonds.

Diamond possesses many interesting properties in addition to its supreme hardness, lustre and fire. Diamond's attraction to grease and its blue fluorescence under X-rays is exploited in recovering diamond from the host rock. Many diamonds also fluoresce in ultraviolet light, and the variable nature of this fluorescence provides an interesting means of identifying jewellery set with diamonds.

DIAMOND

CC	carbon
CS	cubic
H	10
SG	3.515
RI	2.417

ABOVE Diamond crystals in kimberlite from Kimberley, South Africa (top right), and Siberia (centre), and in a beach conglomerate from Namaqualand (top left)

RIGHT The Murchison Snuff Box, in daylight (*right*) and ultraviolet light (*far right*). The box is set with diamonds, most of which fluoresce in ultraviolet light. It bears a portrait of Tsar Alexander II and was presented to Roderick Murchison by the Tsar in 1867, in appreciation of Sir Murchison's geological work in Russia.

Ruby and Sapphire

The beauty of ruby and sapphire lies in the richness and intensity of their colours. Both gems are varieties of the mineral corundum, which is the hardest gem mineral after diamond; consequently they take a brilliant and lasting polish that adds a glittering lustre to their colours.

Pure corundum is colourless, and the colours of corundum are caused by small amounts of chemical impurities (see below). Chromium gives the rich red of ruby and also causes a red fluorescence which further enhances the colour. Sapphire is the name given to all other gem corundum, but is most closely associated with the blue sapphires that are coloured by iron and titanium. Varying amounts and combinations of iron, titanium and chromium give rise to many other colours. Pinkish-orange sapphires are sometimes called 'padparadscha', a name derived

ABOVE Orange sapphire.

LEFT Sapphires and rubies coloured by traces of metallic impurities.

CORUNDUM

CC	aluminium oxide
CS	trigonal
H	9
SG	3.96–4.05
RI	1.76–1.78
DR	0.008–0.010

from the Sinhalese word meaning lotus-colour. Improvement of colour by various forms of heat-treatment has become common in recent years.

Corundum is an aluminium oxide with trigonal crystal structure. Many rubies occur as tabular (flat, table-like) crystals, while sapphire crystals are commonly barrel-shaped or pyramidal. Many sapphire crystals are multicoloured, most often blue and yellow. Ruby and sapphire are strongly pleochroic, so must be cut carefully to display the best colour.

Many fascinating structures and mineral inclusions occur in ruby and sapphire, and are invaluable in distinguishing natural gems from synthetics. Slender rutile needles cause the shimmering effect known as 'silk', and when parallel to the three horizontal crystal directions may give rise to a star effect (see below). Chevron-shaped growth zones, liquid-filled 'feathers' and crystals of zircon and many other minerals are also common as inclusions (see p. 28). The Mogok area of Myanmar has produced fine rubies for many centuries. The best stones display a colour sometimes described as 'pigeon's blood'. A few are mined from the marble in which they formed, but most are recovered from river gravels. The gem gravels of Sri Lanka are similarly famous for an abundance of sapphires of all colours including the orange-pink padparadscha. Fine cornflower-blue sapphires occur in Kashmir pegmatites, but few are mined today. Australia is the present most prolific source of dark blue and golden sapphires. Basaltic rocks are the hosts of both these and the rubies and sapphires of Thailand and Cambodia, but almost all are mined from river gravels. Recent finds of gem gravels in Madagascar include large fine-coloured blue sapphires. Rubies of a good red colour have also been discovered in Kenya, Tanzania and Zimbabwe.

ABOVE Crystal fragments of ruby from Myanmar and sapphire from Kashmir, India.

LEFT Six-rayed star ruby.

RIGHT Faceted sapphire mounted in a turban jewel.

Emerald and Aquamarine

The beauty of these gems, which are varieties of the mineral beryl, lies in their superb colours. Although beryls lack the fire, high lustre and great hardness of diamond and corundum, the unique velvety green of fine emeralds places them among the most precious of gems.

As with many other gems, the colours in beryl are caused by chemical impurities. Pure beryl is a colourless beryllium aluminium silicate, but minute traces of chromium are sufficient to cause the rich green of emerald. Iron gives rise to the greenish-blues of aquamarine and the golden-yellow of heliodor, while pink morganite and the rare red beryl are coloured by manganese. Much greenish-blue aquamarine is heated to produce the blue colour so popular in modern jewellery.

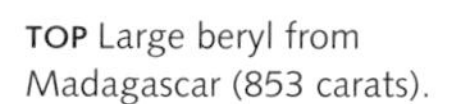

TOP Large beryl from Madagascar (853 carats).

ABOVE Fine yellow beryl (136 carats).

LEFT Beautifully formed crystals of aquamarine, heliodor, morganite and emerald.

Flawless emeralds are exceedingly rare. Many emeralds are fractured and most contain mineral inclusions which, as in corundum, are invaluable in distinguishing natural gems from synthetics. Some inclusions are typical of a particular locality, such as the 'three-phase' inclusions found in Colombian emeralds (see p. 28). By contrast, many aquamarines (blue), heliodors (yellow) and morganites (pink) are virtually flawless; the most common inclusions are slender parallel tubes known as 'rain'.

These variations reflect the origins of the beryl varieties. Aquamarine, heliodor and morganite occur in granites and pegmatites, often as large, beautifully formed hexagonal crystals (see below).

BERYL	
CC	beryllium aluminium silicate
CS	hexagonal
H	7.5
SG	2.63–2.91
RI	1.568–1.602
DR	0.004–0.010

RIGHT Aquamarine.

BELOW Well-formed hexagonal crystals of emerald on calcite.

Brazil is the most abundant source of aquamarine, but fine beryls are also found in Madagascar, California (USA), the Ural Mountains and Adun Chilon in Russia, and many other localities.

The finest emeralds are found around Muzo and Chivor in Colombia, where they occur in veins within dark shales and limestones. This source was mined by the ancient indigenous civilizations and was the origin of the large emeralds that came into Europe and Asia during the 16th century, after the Spanish conquest. Before this time, Egypt, Asia and Austria supplied the emeralds used by the Romans and in medieval Europe. Mica schists are the host rocks for these and many recently discovered deposits, and most emeralds from these localities contain flakes of mica. Few emeralds can match the colour of the best Colombian stones, but some fine emeralds are mined at Sandawana in Zimbabwe, Kitwe in Zambia and Swat in Pakistan.

Opal

Opal used for jewellery falls within one of two categories, precious opal and common opal. Precious opal displays the rainbow iridescence so highly prized in jewellery from Roman times onwards and has a white to dark body colour; common opal has a strong attractive body colour and no iridescence.

The play of colour (seen as flashes of the colours of the rainbow) is caused by the reflection and scattering of light from the minute, uniformly sized and closely packed silica spheres that make up precious opal. This is not a crystalline structure, for opal is one of the few gem minerals that is non-crystalline, or only poorly so.

Common opal shows a wide range of colours and patterns but displays no iridescence because it lacks the regular fine structure of precious opal.

BELOW LEFT Opal matrix from Queensland, Australia.

BELOW Opal in sandstone matrix.

The body colour of precious opal varies, from pale in white opal and black, grey or brown in black opal. Fire opal ranges from yellow to orange and red, and may or may not show iridescence. Water opal is clear and virtually colourless, but with an internal play of colour. In matrix opal, the precious opal fills pore spaces or occurs as narrow veins in the host rock.

The name opal probably derives from 'upala', the Sanskrit word for precious stone. Although this suggests another source for opal, those used by the Romans came from Dubnik near Presov, Slovakia. The Aztecs mined Central American opal and some fine pieces were sent back to Europe by the Spanish Conquistadors.

The discovery of superior quality opal in Australia during the 1870s led to the decline of

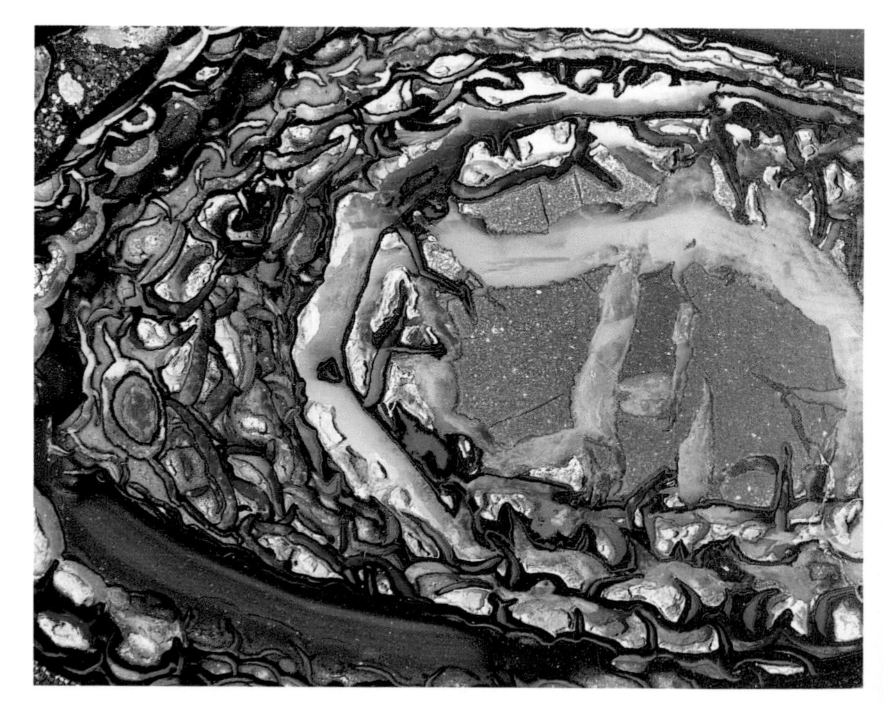

OPAL

CC	silica with some water
CS	amorphous or only poorly crystalline
H	5.5–6.5
SG	1.98–2.25
RI	1.43–1.47

European production. Australia is still the principal source of black and white opal. Mexico produces fine fire opal and water opal, and Brazil is also a significant producer.

Some precious opal forms in gas cavities in volcanic rocks, as in Mexico and Slovakia, but most Australian deposits occur in sedimentary rocks. Opal contains varying amounts of water, and some material may crack if allowed to dry out too rapidly after being mined.

Opal is usually cut as cabochons. Opal veins are often thin, and they may be cut with a backing of the natural rock matrix. Alternatively fragile slivers of vein may be backed with common opal or glass and so form doublets. In triplets the opal is further protected by a capping of clear quartz or plastic.

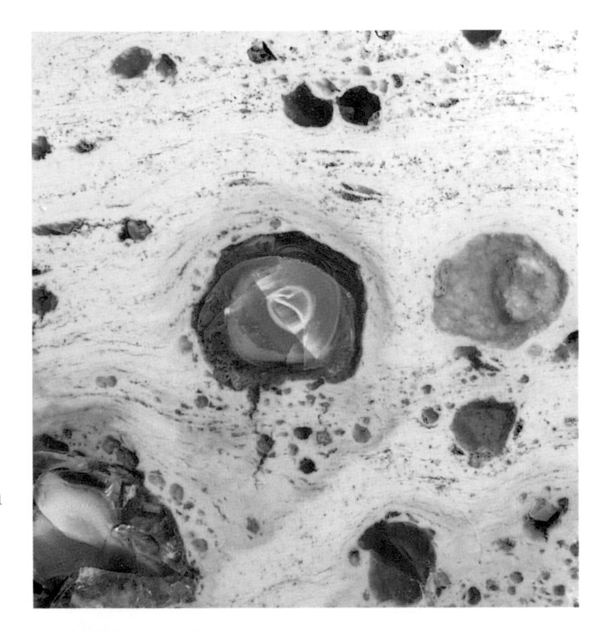

ABOVE RIGHT Fire opal in an alunite matrix.

BELOW RIGHT Necklace of precious opals from Australia.

BELOW Mexican fire opals.

Amethyst and Citrine

Amethyst and citrine are both varieties of quartz, one of the most common minerals in the Earth's crust. The natural abundance and infinite variety of quartz have made it the most widely used of all gem minerals.

Colourless, transparent rock crystal is the purest form of quartz. The colours of amethyst, citrine, rose and smoky quartz are caused by chemical impurities, namely iron in amethyst and citrine, titanium and iron in rose quartz, and aluminium in smoky quartz. Brazil is the most important source of these varieties, which occur as individual crystals or coarsely granular aggregates. In some quartz the colour may be altered by heating, or by irradiation with gamma rays. Natural yellow citrine is comparatively rare and most commercial citrine has been made by heating amethyst. Heat-treated citrine is sometimes referred to as 'burnt amethyst'.

The beauty of much quartz derives from other minerals trapped or grown within it, such as slender golden rutile crystals and tree-like metallic oxides. Some rose quartz contains abundant tiny rutile needles which cause a star effect, usually best seen when a point source of light shines on a cabochon. Asbestos fibres reflect the cat's-eyes seen in some quartz, while the coloured spangles in aventurine quartz are small reflecting flakes of

LEFT Box set with a large citrine.

RIGHT Citrine and amethyst.

green fuchsite mica, brown iron oxides or silvery coloured pyrite crystals.

Tiger's-eye and hawk's-eye (below right) form when blue crocidolite asbestos is replaced by quartz. The asbestos breaks down, either leaving a residue of brown iron oxides, as in tiger's-eye, or retaining the original blue colour, as in hawk's-eye. South Africa produces most tiger's-eye and hawk's-eye.

Fine transparent amethyst, citrine and smoky quartz is generally faceted, but other qualities and varieties are cut as cabochons. Quartz is quite tough so that most varieties can be carved, and rock crystal has been fashioned into delicate bowls for many centuries.

QUARTZ	
CC	silica
CS	trigonal
H	7
SG	2.65
RI	1.544–1.553
DR	0.009

RIGHT Purple amethyst crystals lining a geode.

BELOW RIGHT Tiger's-eye and blue hawk's-eye.

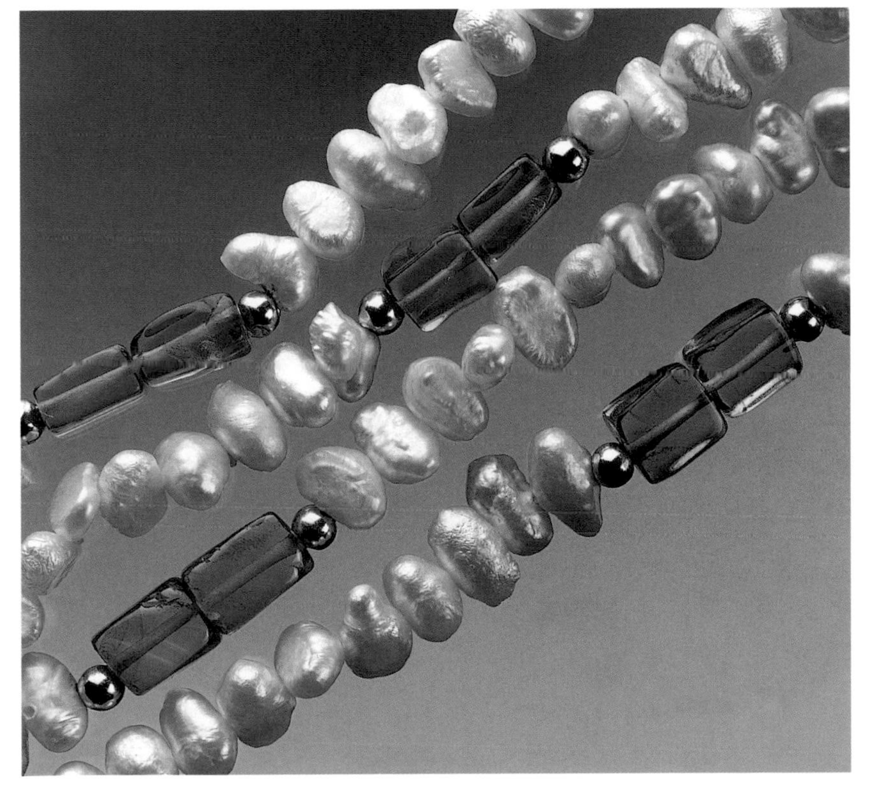

BELOW Necklace of amethyst with baroque cultured pearls.

Garnet

The name garnet may derive from the Latin 'granatum', meaning pomegranate. The red colours of the pomegranate flesh are similar to those of many garnets. Garnet is a group name for the silicate minerals almandine, pyrope, spessartine, grossular, andradite and uvarovite, so is a far more diverse gem than its name suggests. The garnet minerals share a similar cubic crystal structure and are related in chemical composition. Gem-quality garnet occurs in many countries, and beautifully formed crystals have been prized as gems for over 5000 years.

Red almandine and pyrope are the most widely used of all garnets. Almandine cabochons and carved stones have been popular since Roman times, and much 19th-century jewellery was set with pyrope garnets from an area now in the Czech Republic.

Almandine is an iron aluminium garnet and pyrope a magnesium aluminium garnet, but the iron and magnesium can replace each other and so give rise to a continuous gradation of iron magnesium aluminium garnets known as the almandine–pyrope series. This chemical variation causes a gradation in gem properties, so that two apparently similar red garnets may differ

ABOVE Brooch set with almandine garnets.

BELOW LEFT Pyrope garnet.

BELOW Faceted garnets showing a range of colours.

GARNET

CC	silicates of magnesium, iron, calcium manganese and aluminium
CS	cubic
H	6.5–7.5
SG	3.58–4.32
RI	1.714–1.887

ABOVE LEFT Spessartine garnet.

RIGHT Six-rayed star garnet, from Australia (see p. 53).

BELOW Green demantoid garnet in ring, from the collection of Sir Arthur Herbert Church.

in density and refractive index. Other garnet minerals form similar series. Iron causes the brownish and purplish–reds of almandine, but blood-red pyrope is coloured by traces of chromium.

Spessartine and hessonite are orange garnets. Spessartine is manganese aluminium garnet and hessonite is a variety of grossular, the calcium aluminium garnet. Pure grossular is colourless. Iron causes the orange of hessonite and vanadium colours the green Kenyan grossular garnet called tsavorite. Until the discovery of the brilliant green crystals of tsavorite at Tsavo in Kenya in 1974, hessonite was the most important grossular gem.

Demantoid is a rare, vivid green variety of andradite, the calcium iron garnet. It is more fiery than diamond, but this effect may be masked by the glorious colour, which is caused by chromium impurities. The demantoid variety is softer than other garnets.

Garnets may contain distinctive inclusions, such as the rutile or hornblende needles that cause a star effect in some almandine. Much demantoid contains groups of fibrous minerals clustered like 'horsetails', while most hessonite can be identified by its granular and 'treacly' texture.

Tourmaline

TOURMALINE	
CC	complex borosilicate
CS	trigonal
H	7–7.5
SG	3.0–3.25
RI	1.610–1.675
DR	0.014–0.034

Tourmaline shows the greatest colour range of any gemstone, although pink and green stones are the most popular. Some crystals are multicoloured, with different colours occurring at either end of the crystal (below), or forming a core and rim, as in 'water melon' tourmaline.

Much tourmaline occurs as beautifully formed, elongate crystals with a distinctive 'rounded triangular' shape in cross section. Many crystals exhibit polarity – that is, the colour, electrical properties and the crystal forms are different at either end of the crystal. These fascinating variations arise from the complex crystal structure and chemistry of tourmaline.

Tourmaline is a borosilicate mineral that varies greatly in its composition. John Ruskin, the eminent Victorian thinker, wrote of tourmaline that "the chemistry of it is more like a medieval doctor's prescription than the making of a respectable mineral". Although most gem tourmaline is lithium-rich, there is no simple correlation between chemical composition and colour.

Tourmaline is strongly pleochroic and the deepest colour is always seen when looking down the length of the crystal. Some green and blue tourmaline appears almost black in this direction, so that it is important to position rough material correctly when cutting gems. Most tourmaline forms in granites and pegmatites. Fine gems come from many localities worldwide, but some of the more famous include Minas Gerais and Paraiba in Brazil, the Ural Mountains in Russia, and California, USA.

ABOVE Faceted tourmaline showing colour range.

LEFT Pink and green tourmaline crystal from California, USA.

BELOW Chinese carving in pink rubellite tourmaline.

Topaz

In the past it was thought by many people that all yellow gems were topaz and that all topaz was yellow. Topaz, however, varies from pale blue and colourless to yellow, orange, brown and pink. The pink stones so popular in Victorian jewellery were produced by heating golden-brown topaz from Ouro Preto, Brazil. Today, vivid blue topaz is produced by irradiating and then heating certain colourless material (bottom right).

Topaz is reputedly named from Topazius, the ancient Greek name for Zebirget in the Red Sea. This may or may not be true however, as the gem mined from this island is the green peridot. The modern topaz more probably equates in history with the stone described by Pliny as chrysolite.

Topaz is an aluminium silicate that contains up to 20% fluorine or water, its physical and optical properties varying according to the proportions of water and fluorine present. Golden-brown and pink topaz contains more water and tends to form longer crystals. Such topaz is less dense and more highly refractive than the fluorine-rich colourless, pale blue and yellow topaz which occur as squat crystals. Although topaz is the hardest silicate gem mineral, it cleaves easily along a direction parallel to the base of the crystal.

Topaz occurs chiefly in granites and pegmatites, and in the contact zones around them. Some gem-quality crystals are huge and crystals weighing many kilograms are not uncommon. Many mineral collections have fine examples of golden and pink crystals from Ouro Preto, Brazil, blue crystals from the Ural Mountains and yellow topaz from Saxony, Germany.

TOPAZ

CC	aluminium fluorosilicate with some hydroxyl
CS	orthorhombic
H	8
SG	3.49–3.57
RI	1.606–1.644
DR	0.008–0.011

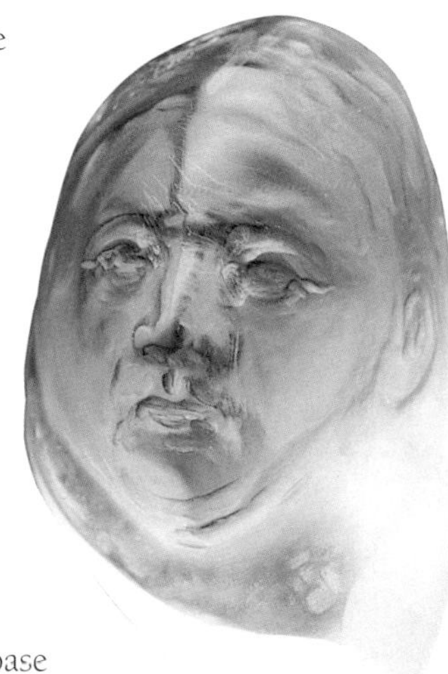

RIGHT Image of Queen Victoria carved in topaz.

BELOW RIGHT Blue irradiated topaz pebbles and faceted topaz.

BELOW Topaz crystals and faceted topaz showing colour range.

Chrysoberyl

There are three beautiful gem varieties of chrysoberyl. The pale yellow-green stones from Brazil show an extraordinary brilliance, popular in Spanish and Portuguese jewellery of the 18th century. Some chrysoberyl contains numerous parallel, needle-like inclusions and, when cut as cabochons, displays the sharpest of cat's-eyes (right). This chrysoberyl variety is known simply as cat's-eye or as cymophane. It has been said that the third variety, alexandrite was discovered in the Ural Mountains in 1830 on the birthday of Czar Alexander II . Alexandrite is famous for its colour change, from deep green in daylight to red in tungsten light.

Chrysoberyl is a beryllium aluminium oxide, exceeded in hardness only by diamond and corundum. The yellow, green and brown colours are caused by small amounts of iron or chromium, which also causes the alexandrite effect. In addition to the change in colour with different lighting conditions, alexandrite is remarkably pleochroic: it appears red, orange-yellow and green when viewed from different crystal directions.

Synthetic corundum and spinel have been made as poor imitations of alexandrite. Synthetic chrysoberyl is also made. Chrysoberyl crystallizes in the orthorhombic system. Simple crystals are rare, but fine trillings or twinned crystals, which appear hexagonal at first glance, are found in Brazil and the Ural Mountains (left). Chrysoberyl forms in beryllium pegmatites, but most gem-quality material is mined from river gravels in Sri Lanka, southern India, Brazil and Myanmar.

CHRYSOBERYL	
CC	beryllium aluminium oxide
CS	orthorhombic
H	8.5
SG	3.68–3.78
RI	1.742–1.757
DR	0.009

RIGHT Chrysoberyl cut as a cabochon to show the bright line with the appearance of the eye of a cat, a cat's-eye.

BELOW LEFT A twinned Alexandrite crystal.

BELOW RIGHT Faceted yellow chrysoberyl.

Peridot

Peridot is the transparent gem variety of the mineral olivine. Peridot's attraction is in its colour, ideally a rich 'oily' green but ranging from pale golden-green to brownish-green. Peridot is a French word, possibly derived from the Arabic 'faridat', meaning gem. The Red Sea island of Zebirget (St John's Island) was the source of fine dark green peridot used by the ancient Mediterranean civilizations. The Greeks and Romans knew the island as Topazios and called this green gem topaz. Today peridot is found in Myanmar, Norway, Arizona (USA), Hawaii, the Canary Islands, Australia and South Africa.

Peridot is a magnesium iron silicate mineral that occurs in basalts and peridotite rocks. Iron causes the colour of peridot and the proportion of iron present determines the shade and depth of colour. The pale golden-green and most valuable deep green peridots contain smaller amounts of iron than the less attractive brownish-green stones.

Peridot is relatively soft and has a distinctive oily lustre. It is strongly doubly refractive so that the facet edges of a cut stone appear doubled when viewed through the stone. Many peridots contain inclusions that resemble water-lily leaves (see p. 28).

In 1952 many brown stones thought to be peridots were found instead to be a borate mineral (see p. 5). This newly recognized mineral was named sinhalite after its most important source, Sri Lanka, for which the Sanskrit name is Sinhala.

PERIDOT	
CC	magnesium iron silicate
CS	orthorhombic
H	6.5–7
SG	3.22–3.40
RI	1.635–1.695
DR	0.036

LEFT Fine cut peridot of 146 carats.

BELOW RIGHT Crystal of peridot from Zebirget (St John's Island), Red Sea.

BELOW LEFT A rough mass of olivine.

BELOW Carved peridot set in a ring.

Zircon

Although the name zircon comes from the Arabic 'zargoon', meaning vermilion or golden coloured, zircons also occur in a wide range of subtle greens and browns and are occasionally colourless. Such stones have been used for many centuries in Indian and Sri Lankan jewellery.

In jewellery, zircons are most familiar as lustrous and fiery vivid blue, golden or colourless stones, which are usually cut as round brilliants. These colours are produced artificially by heating brown zircon from Thailand, Cambodia and Vietnam. Heating in an oxygen-free atmosphere produces blue zircon, which may then be heated in air to give the golden colour; both processes produce some colourless material. Heat-induced colours sometimes fade on exposure to light but may be restored by careful reheating.

Zircon resembles diamond with its fine lustre and fire, so that colourless stones have been both mistakenly identified as diamonds and purposely used as diamond simulants. But, zircon is tetragonal in structure and can be distinguished from diamond by its double refraction. Although zircon is moderately hard it can be very brittle, and the facet edges of cut stones are easily chipped.

Most zircons contain traces of uranium or thorium, or rare-earth elements, which produce spectra with abundant absorption lines. Radiation from these elements may gradually disrupt and transform the crystalline structure to a 'metamict' state and such zircons are usually green and cloudy with lower refractive indices, density and hardness.

ZIRCON	
CC	zirconium silicate
CS	tetragonal
H	7.5
SG	4.6–4.7
RI	1.923–2.015
DR	0.042–0.065

BELOW LEFT Fine tetragonal zircon crystals from Norway.

BELOW Natural brown and heat-treated blue zircons.

RIGHT Faceted zircon showing its natural colour range.

Spinel

During medieval times, magnificent red gems known as 'Balas rubies' found their way into many royal treasuries. These stones are in fact spinels, and one of the most famous, the Black Prince's Ruby, is set in the British Imperial State Crown (see below). The old name is probably derived from Balascia (now Badakhshan) in Afghanistan, the supposed source of these gems.

The only difference in composition between spinel and ruby is that ruby is aluminium oxide and spinel is magnesium aluminium oxide. However, spinels are more variable because the magnesium can be replaced by iron, manganese or zinc, and chromium and iron may replace the aluminium. These variations give rise to a range of colours (see below right) and cause differences in density and optical properties. Pure spinel is colourless; reds and pinks are because of a small amount of chromium, iron causes green and blue, and zinc spinel is blue.

Spinel crystallizes in the cubic system. Well-formed octahedral crystals are common, as are flat spinel twins or macles, which resemble diamond macles in shape (see p. 35). Most gem-quality crystals come from thermally altered limestones that are rich in dolomite. Some are mined directly from such rocks, but most come from the richer concentrations in gravels in Sri Lanka, Myanmar and Brazil.

Some spinels contain planes of tiny octahedral crystals, while others contain parallel sets of rutile needles which give rise to a star effect in cabochon-cut stones. These spinels, like some garnets, may show both six-rayed and four-rayed stars.

SPINEL	
CC	magnesium aluminium oxide
CS	cubic
H	8
SG	3.58–4.06
RI	1.714–1.750

LEFT Octahedral crystals of red spinel.

BELOW LEFT Large red spinel (called The Black Prince's Ruby) set in the Imperial State Crown. A small ruby is set in the spinel.

BELOW Faceted spinel showing a range of colour.

Tanzanite

Tanzanite, discovered in Tanzania in 1967, is a blue variety of the mineral zoisite. It is strongly pleochroic, displaying rich blue, magenta and yellowish-grey colours when viewed from different angles. The crystals, which occur in various colours including blue, green, yellow and brown, are heated to change their colour to the more attractive blue colour. Care must be taken when setting or handling tanzanite as it has one perfect cleavage, and the use of ultrasonic cleaners should be avoided as this may result in the stone shattering. The pink variety of zoisite is called thulite after Thule, an old name for Norway. Thulite is opaque and carved or polished for use as a decorative stone, beads or cabochons. Anyolite is an attractive ornamental rock comprising green zoisite surrounding included ruby crystals.

TANZANITE	
CC	calcium aluminium hydroxysilicate
CS	orthorhombic
H	6
SG	3.38
RI	1.696–1.702
DR	0.007

TAAFFEITE	
CC	beryllium magnesium aluminium oxide
CS	hexagonal
H	8
SG	3.613
RI	1.718–1.723
DR	0.004

LEFT Crystal and faceted tanzanite, heat-treated to improve the blue colour.

BELOW Tanzanite ring.

RIGHT Faceted taaffeite.

Taaffeite

The gemstone illustrated below is of the first taaffeite ever found. It was discovered by Count Taaffe of Dublin in 1945, among gems broken out of old jewellery. Taaffeite is a very rare beryllium magnesium aluminium oxide mineral, which resembles spinel in colour. It was because of this resemblance that the first taaffeite had already been cut as a gemstone before being identified as a new mineral, the first time such an event had happened.

Count Taaffe recognized that the gem could not be a spinel as spinel is singly refractive and the gem he had found was doubly refractive (see p. 17). The stone was sent to London, to the Precious Stone Laboratory and then to The Natural History Museum, where a small slice was taken off the back of the stone (the culet) for X-ray analysis. It was found to be a new mineral. Since then many more specimens have been found, which include material varying from pale mauve to a sapphire blue, and a ruby-red coloured taaffeite from Sri Lanka.

In 1999, a similar event occurred, when a 14.02 carat gemstone was also tested and found to be johachidolite, the first gem of a mineral only known previously as tiny grains less than 1 mm across.

Benitoite

Benitoite was discovered in 1906 near the San Benito River in California (USA), and this remains the only known locality for this barium titanium silicate mineral. There is more than one version of the events that led to the discovery of benitoite. The generally accepted version is that a prospector was looking for mercury and copper minerals near the San Benito River when he found small blue crystals that he thought were sapphires. They eventually reached G.D. Louderback at the University of California, who recognized them as a new mineral and named them benitoite after the locality.

Benitoite shows a strong dichroism from blue to colourless as the stone is turned. The best blue colour is seen when looking through the side of the tabular crystals rather than through from top to bottom. The size of the cut stones is therefore limited by the shape of the crystal, as they need to be cut with the table facet parallel to the short sides of the crystal. Another feature of benitoite is its exceptionally strong dispersion. Its fire is similar to that of diamond, but is masked by the blue colour. Colourless benitoite and pink benitoite have also been found from the same locality.

BENITOITE	
CC	barium titanium silicate
CS	hexagonal
H	6.5
SG	3.65–3.68
RI	1.757–1.804
DR	0.047

IOLITE	
CC	magnesium aluminium silicate
CS	orthorhombic
H	7–7.5
SG	2.57–2.61
RI	1.53–1.55
DR	0.008–0.012

LEFT Faceted benitoite.

RIGHT The colour of the iolite cube varies when viewed from different directions because it is pleochroic.

Iolite (cordierite)

Iolite is a variety of the mineral cordierite. The name iolite is derived from a Greek word meaning violet, after its colour. The mineral cordierite is named after the French geologist Cordier. Iolite is famous for its pleochroism, appearing intense blue in one direction, yellowish-grey or blue in another, and becoming almost colourless as the stone is turned to the third direction (below). By holding an iolite up to the sky and estimating the position of the Sun, this feature may well have been used as an aid to navigation by the Vikings.

Iolite does occur as well-formed crystals, but is more often found as rolled pebbles in the gem gravels of Sri Lanka, Myanmar and Madagascar. Other localities include India and the Northwest Territories of Canada. African localities include Namibia and Tanzania.

Andalusite, Fibrolite (sillimanite) and Kyanite

Andalusite, fibrolite and kyanite share the same chemistry – aluminium silicate – but differ in origin and crystal structure. Andalusite shows spectacular red and green pleochroic colours, both visible in the fine stone illustrated (below).

Andalusite is named after the province Andalusia in southern Spain where the mineral was first discovered. Other localities where andalusite is found, mainly as water-worn pebbles, include Sri Lanka and Brazil. Also, very attractive blue stones have been found in Myanmar, and greyish-white material from the gem gravels of Sri Lanka has been cut as cabochons to show a cat's-eye.

Fibrolite is a rare variety of the mineral sillimanite (named after Silliman of Yale University, USA). Stones from Myanmar show bluish-violet and pale yellow pleochroic colours. Fibrolite cleaves easily so large cut stones are very rare.

Kyanite forms blue or green flat-bladed crystals, which are seldom used as gems although they may be cut as cabochons and show a cat's-eye. Kyanite is extraordinary in that its hardness in one direction is quite different to that in another. Along the length of the crystals, the hardness may be as low as 5 on Mohs' scale. The fibrous nature of the crystals and the cleavage make it liable to break along its length. Across the crystal the hardness is greater, with values as high as 7 on Mohs' scale. Kyanite crystals are often colour zoned, with darker colours towards the centre of the crystal, and may be white or colourless towards the edges. Blue kyanite shows a marked pleochroism, appearing blue, violet-blue and colourless when viewed from different directions. Localities for gem-quality kyanite include Myanmar, India and Kenya.

ANDALUSITE

CC	aluminium silicate
CS	orthorhombic
H	7.5
SG	3.15–3.17
RI	1.634–1.641
DR	0.007–0.011

FIBROLITE

CC	aluminium silicate
CS	orthorhombic
H	7.5 (crystals), 6–7 (fibrous material)
SG	3.14–3.18
RI	1.658–1.678
DR	0.020

KYANITE

CC	aluminium silicate
CS	triclinic
H	7 (across crystal), 5 (along crystal)
SG	3.65–3.69
RI	1.715–1.732
DR	0.017

FAR LEFT Faceted andalusite.

LEFT Faceted fibrolite (sillimanite).

Spodumene

Kunzite and hiddenite are the pink and green varieties of the mineral spodumene. Kunzite is distinctly pleochroic and cleaves easily along two crystal directions, so is difficult to cut. The colours seen when the gemstone is turned are pink to violet, deep pink to deep violet, and colourless and these may fade in daylight. Kunzite is named after G.F. Kunz, an American gemmologist. Another variety of spodumene is the green hiddenite, coloured by chromium. Hiddenite was named after Hidden, a superintendent at the mining company in North Carolina where this new variety was discovered in 1879. It is also strongly pleochroic, showing green, bluish-green and yellowish-green when viewed from different directions.

The first gem-quality spodumene was yellowish-green and found in about 1877 in Brazil, where it was sold as chrysoberyl. Two years later, green hiddenite was discovered in North Carolina and lilac-coloured (or pink) kunzite in Connecticut, USA. Gem-quality kunzite was found in 1902 in California. Kunzite and hiddenite are also known from Madagascar and Myanmar, where blue spodumene has also been recorded.

Spodumene

CC	lithium aluminium silicate
CS	monoclinic
H	7
SG	3.17–3.19
RI	1.660–1.675
DR	0.015

ABOVE Strongly dichroic pink kunzite crystal and faceted kunzite.

LEFT Pale spodumene with faceted hiddenite.

Kornerupine

Kornerupine is a rare borosilicate mineral, which occurs in a range of greens and browns – this emerald-green colour is exceptional. Lorenzen first discovered kornerupine in Greenland in 1884, and named it after Kornerup, a Danish scientist. Gem-quality material was not found until 1912. The greenish-brown material is strongly pleochroic and looks green or reddish-brown when viewed from different directions. The best colour is seen through the side of the crystal, so the gem-quality material is cut with the length of the crystal parallel to the table facet.

Localities include the gem gravels of Myanmar, Sri Lanka and Madagascar. Some material from Sri Lanka and Africa can be cut as cabochons to show a cat's-eye. African localities of the green gem-quality material include Kenya and Tanzania.

KORNERUPINE	
CC	magnesium iron aluminium borosilicate
CS	orthorhombic
H	6.5
SG	3.27–3.45
RI	1.665–1.680
DR	0.013

SCAPOLITE	
CC	complex aluminium silicate (isomorphous series)
CS	tetragonal
H	6
SG	2.60–2.71
RI	1.540–1.577
DR	0.009–0.020

LEFT Faceted kornerupine.

Scapolite

Scapolite of gem quality was first found in Myanmar. Scapolite resembles feldspar in composition and is made up of a series of chemical compositions, from the sodium-rich marialite at one end of the series to the calcium-rich mineral called meionite at the other. The gemmological properties vary with composition, with the sodium-rich members having the lower values.

Scapolite is distinctly pleochroic. The violet stones show a dark or light blue and violet and the yellow stones show pale yellow and colourless when viewed from different directions. Scapolite is seldom cut, except for collectors, as there are three directions of easy cleavage, which makes the stone vulnerable and difficult to work.

In 1913 the first gem-quality scapolite was found in Myanmar. When cut as cabochons, this fibrous white, pink and purple material showed a cat's-eye effect. Transparent yellow scapolite has been found in Myanmar, Madagascar and Brazil. African localities include Mozambique and Kenya.

RIGHT Faceted scapolite from Myanmar.

Diopside and Enstatite

Although diopside and enstatite are fairly common silicate minerals, they are rare as gems. The most appealing gem varieties are the dark brown star diopsides and the rich green chrome diopsides and chrome enstatites. Diopside also occurs as dark bottle-green, brown, blue and colourless stones. High iron content causes the darker colours and increased densities. Fibrous or cloudy material of both minerals can be cut as cabochons to show a cat's-eye or star effect.

Chrome diopside has been found in Myanmar, the Kimberley diamond mines, Siberia and the Hunza Valley of Pakistan. Gem-quality diopside is found in Austria and Italy. Other localities include the gem gravels of Sri Lanka and Brazil, and regions of Canada and the USA. Enstatite occurs as emerald green, brownish-green or colourless crystals, but it is more usually found as rolled pebbles in gem gravels. It is found in areas where diamonds are mined, such as South Africa and also in a variety of rock types, including kimberlite, and so appears in many localities worldwide.

DIOPSIDE

CC	calcium magnesium silicate
CS	monoclinic
H	5.5
SG	3.29 plus
RI	1.660–1.701
DR	0.030

ENSTATITE

CC	magnesium iron silicate
CS	orthorhombic
H	5.5
SG	3.26–3.28
RI	1.663–1.673
DR	0.010

LEFT Faceted diopside

Sphene (titanite)

Sphene is prized for its great lustre and fire, but is rather soft. Gem-quality crystals of this calcium titanium silicate range in colour from yellow and golden-brown to greens and are often twinned. Sphene is cut as a brilliant cut or mixed cut in order to show the exceptional fire to best effect. The dispersion is greater than that of diamond. High birefringence is seen as a doubling of the back facets when viewed through the front of the stone. Pleochroism is also distinct. The three colours seen when the stone is turned are near colourless, greenish-yellow and reddish- or brownish-yellow.

Localities include the Austrian and Swiss Alps, Ontario (Canada), Madagascar, Sri Lanka, Myanmar, California (USA) and Brazil.

SPHENE

CC	calcium titanium silicate
CS	monoclinic
H	5.5
SG	3.52–3.54
RI	1.885–2.050
DR	0.105 to 0.135

RIGHT Faceted sphene.

Chalcedony and Jasper

The attraction of the more subtly coloured chalcedonies and jaspers lies in the enormous variety of patterns and textures that develop as these minerals grow, the growth bands and trapped mineral fragments often resembling exotic landscapes and gardens.

Chalcedony (including agates) and jasper are the chameleons of the gem world, displaying a bewildering variety of subtle colours and patterns. Like amethyst and citrine, they are varieties of quartz, but differ in that they consist of minute fibres and grains that are visible only at high magnification. The two groups are distinguished by their internal structure.

Chalcedony, which includes all agates, carnelian and chrysoprase, forms by the build-up of thin layers of tiny quartz fibres. This layered structure is visible as the colour banding in agate and onyx. Because of its fibrous structure, chalcedony is extremely tough and has been recognized for centuries as an excellent carving material. This toughness and colour banding are most brilliantly exploited in the classical Roman cameos (see p. 29).

Pure chalcedony is a translucent grey or white, while coloured varieties and patterns are the result of variable contents of impurities. Iron oxides cause the browns in agates and sardonyx, and the red of carnelian. Apple-green chrysoprase is coloured by nickel (see below left), while the darker greens of plasma and prase are the result of countless tiny crystals of chlorite and actinolite. Bloodstone is plasma that is speckled with red jasper (left).

Some moss agate contains a tangle of moss-like green minerals, but the variety known as mocha stone is patterned with iron or manganese oxides. The iridescent colours of fire agate are produced by interference of light at thin layers of regularly spaced iron oxide crystals within the chalcedony.

Chalcedony is porous and may be stained with a variety of metallic salts. Natural black onyx is rare and most onyx in jewellery is produced by soaking agate in sugar solution, then heating it in sulphuric acid to carbonize the sugar particles. A similar process was used by the Romans.

ABOVE Frog carved in bloodstone, by Paul Dreher.

LEFT Chrysoprase cameo.

BELOW 'Poppy' jasper from Morgan Hill, Santa Clara Co., California.

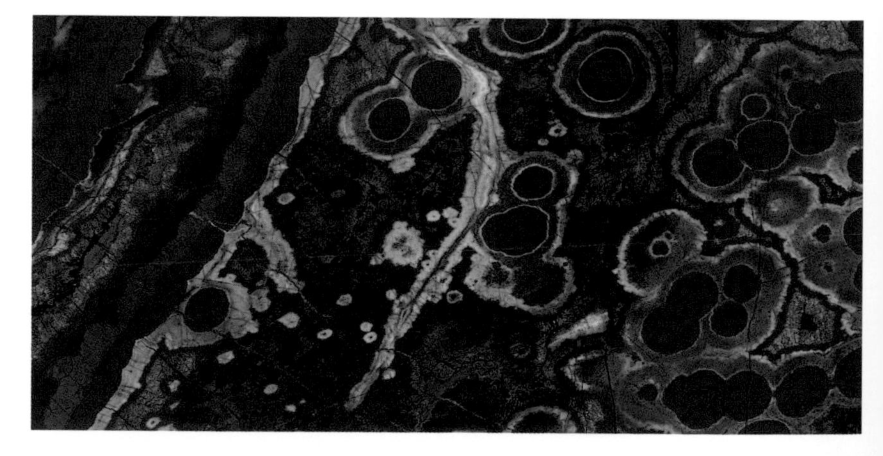

Jasper consists of a mass of minute, randomly arranged, interlocking quartz crystals. It is opaque and contains large amounts of colourful impurities which consist mostly of red and yellow iron oxides or green chlorite and actinolite. Because of this range of colours jasper is used for carvings and in mosaics and inlays.

Chalcedony and jasper occur worldwide, but Brazil and Uruguay are the most important sources of agates.

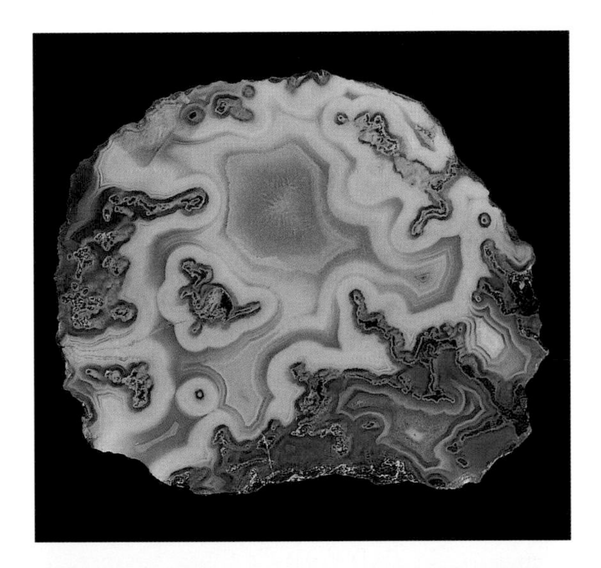

ABOVE Mocha stone with black veins of iron and manganese oxides.

TOP RIGHT Agate from Idar-Oberstein, Germany.

RIGHT Agate from Brazil.

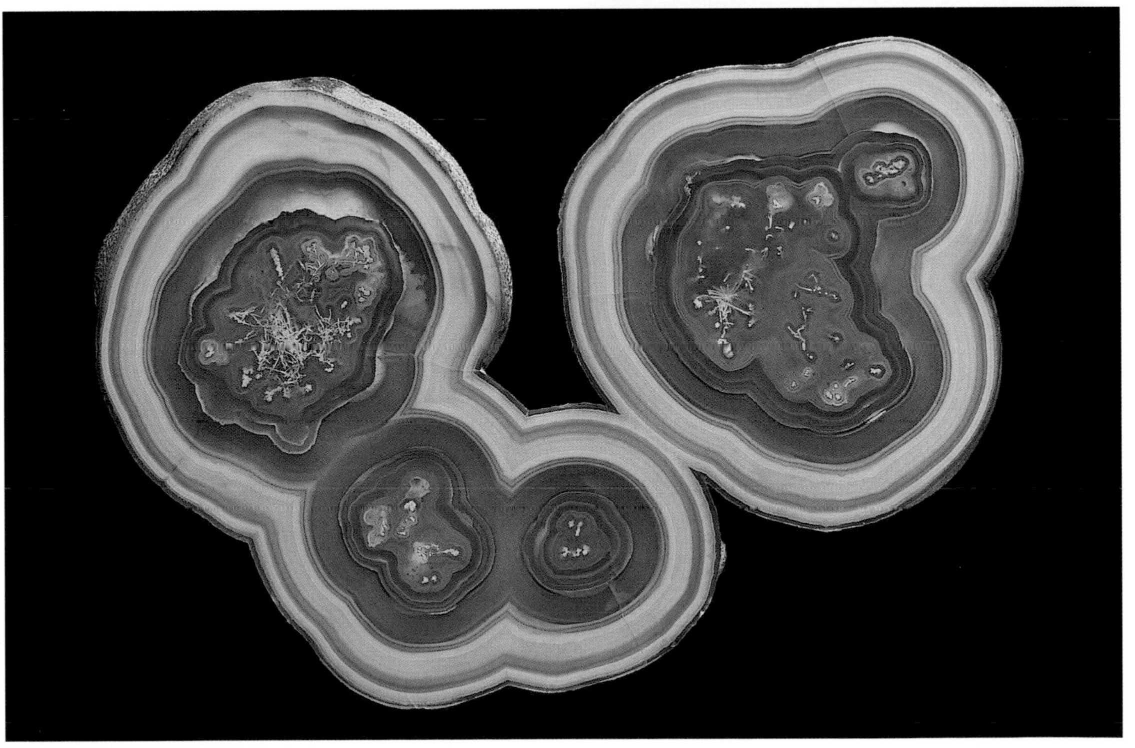

CHALCEDONY AND JASPER

CC	silica
CS	trigonal (microcrystalline)
H	6.5–7
SG	2.6
RI	1.535 (mean)
DR	rarely seen

Jade

Jade is named from the Spanish 'piedra de hijada', a term originally used for the green stone carved by the indigenous civilizations of Central America (see below). In Europe, the name was extended to a stone with similar properties that was used in imported Chinese carvings. It was not until 1863 that the French mineralogist Damour demonstrated that the name jade was being used for nephrite and jadeite, stones composed of two quite different minerals.

Although not particularly hard, nephrite and jadeite are tougher than steel. Consequently they have been used since Neolithic times for weapons and tools, and in later ages for the most delicate carvings. This toughness stems from the structure of jade, which consists of a mass of microscopic, interlocking fibres and grains. The structure also gives rise to the dimpled 'orange peel' surface seen on older carvings. The diamond abrasives used today produce a much smoother finish.

Jadeite is a sodium aluminium silicate. Nephrite is a calcium magnesium aluminium silicate containing variable amounts of iron. Nephrite is generally green to creamy-white, while jadeite is more diverse, varying from white to green, brown, orange or, rarely, lilac. Iron gives rise to most

JADEITE	
CC	sodium aluminium silicate
CS	monoclinic
H	6.5–7
SG	3.3–3.5
RI	1.66 (mean)

ABOVE RIGHT Violet jadeite.

LEFT Jadeite mask, from the ancient Mayan civilization, South America.

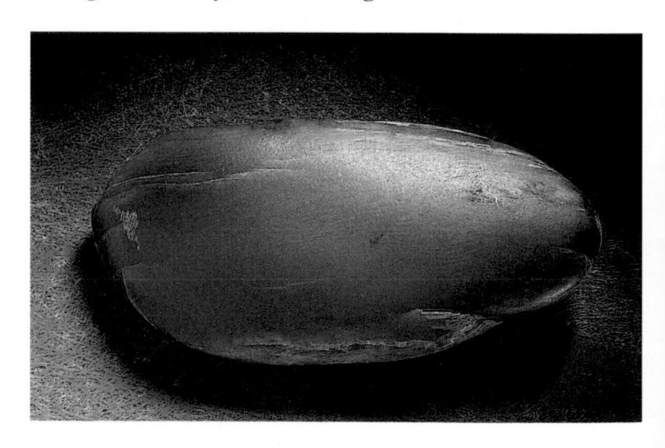

BELOW RIGHT Nephrite.

NEPHRITE

CC	calcium magnesium aluminium silicate, with some iron
CS	monoclinic
H	6.5
SG	2.9–3.1
RI	1.62 (mean)

greens and browns, and manganese is thought to colour lilac jadeite. The most prized of all jades is 'imperial jade', the transparent emerald-green jadeite coloured by chromium. Much jade occurs as water-worn boulders which may develop a weathered brown outer skin.

Jade carving was perfected by the Chinese, yet from about 200 BC nephrite had to be imported from East Turkestan. Jadeite was introduced into China from Myanmar much later, around 1750. Jadeite is the rarer of the jades, and Myanmar remains the only commercial source of jadeite. Much of the Central American jadeite, which was worked from about 1500 BC until the Spanish conquest, originated in Guatemala. Today fine nephrite is mined from several sources. Spinach-green nephrite has been produced from near Irkutsk in Siberia since 1850, and the Maoris have been carving material from the South Island of New Zealand for several centuries. Taiwan and British Columbia are more recent sources.

RIGHT Carved nephrite: a tiki from New Zealand (left), a seal from British Columbia, and a Chinese vase.

ABOVE RIGHT Delicate nephrite bowl.

Turquoise

Turquoise was one of the first gems to be mined and also inspired some of the earliest known imitations. Its colour was greatly admired by the ancient Egyptians and their predecessors, who first mined turquoise in Sinai over 6000 years ago. Demand outstripped supply very early because blue and green glazed steatite imitations have been found in graves dated at 4000 BC.

The finest turquoise occurs in Nishapur, Iran, where it has been mined for about 3000 years. Today most turquoise is produced in the south-western United States, where it is set in fine native American jewellery. This source was known to the Aztecs, who used turquoise in mosaic work on ritual masks and other ornaments (below).

Turquoise is a phosphate, most commonly forms as minute crystals, and occurs chiefly as veins and nodules around copper deposits in arid regions. The sky-blue colour is caused by the copper that is an essential part of its chemical composition. Much turquoise, however, also contains iron, which causes the less favoured greenish tints. Gems patterned with dark iron oxides are cut from 'turquoise matrix', which consists of turquoise and fragments of the host rock. Turquoise is relatively soft and has a waxy lustre. As it is porous, its colour may deteriorate if skin oils and cosmetics are absorbed during wear. The more porous material is friable and sometimes fades as it dries out on exposure to air. Such turquoise is usually 'colour stabilized' and made more durable by bonding with plastic resins or silica.

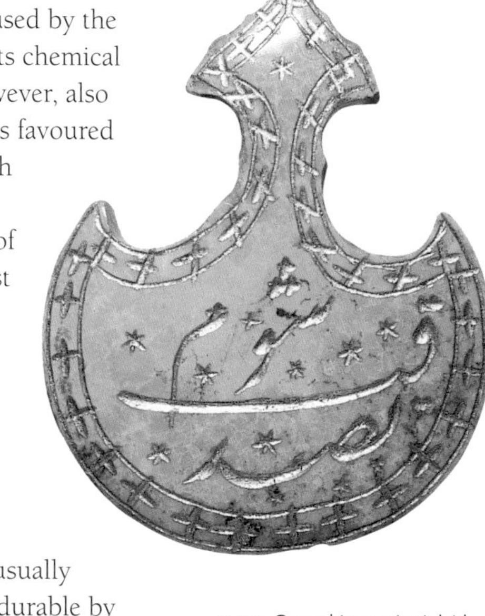

ABOVE Carved turquoise inlaid with gold.

TURQUOISE	
CC	hydrated phosphate of copper and aluminium
CS	triclinic (cryptocrystalline)
H	5–6
SG	2.6–2.9
RI	1.62 (mean)

FAR LEFT Turquoise vein in shale, from Victoria, Australia.

LEFT Mask of Aztec god Quetzalcoatl covered in a mosaic of turquoise pieces.

Malachite and Azurite

Malachite and azurite are vivid green and blue minerals of copper carbonate. Often found as opaque lumps with nobbly surfaces (botryoidal habit), the light green and dark green banding in malachite gives it an attractive and distinctive appearance when polished or carved. Malachite beads are particularly popular, but larger pieces including bowls and carved animals are also made. Although crystals of azurite have been found, they are usually too small to be faceted.

Malachite is often found on its own, but usually azurite is intergrown with malachite rather than alone. Banded azurite and malachite found in Chessy, near Lyons (France), is called chessylite. Malachite and azurite are found in large quantities in copper mining areas such as those in Chile, Australia, Russia, Africa and China. Zaire is the most important producer of malachite.

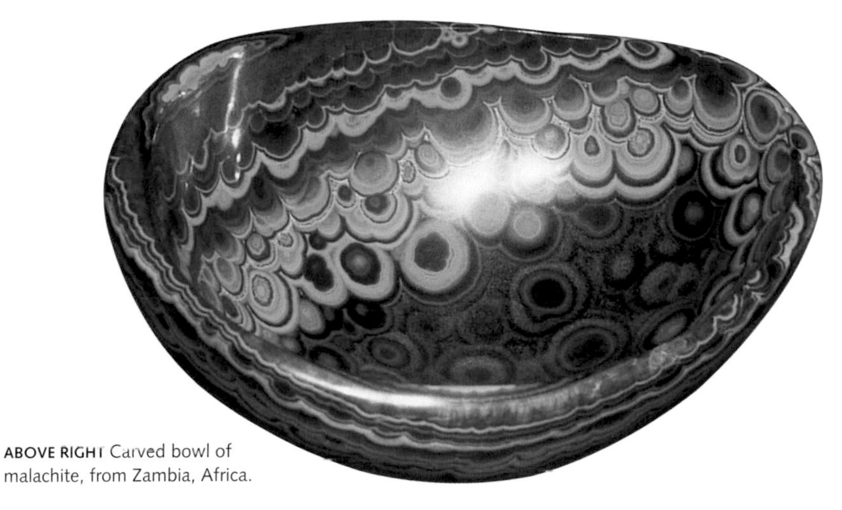

ABOVE RIGHT Carved bowl of malachite, from Zambia, Africa.

BELOW LEFT Malachite from Bisbee, Arizona, USA.

BELOW RIGHT Polished malachite from Gumeschevish, Russia.

MALACHITE

CC	copper hydroxy-carbonate
CS	monoclinic
H	4
SG	3.800
RI	1.85 (mean)

AZURITE

CC	copper hydroxy-carbonate
CS	monoclinic
H	3.5
SG	3.770
RI	1.730–1.740

Lapis lazuli

Lapis lazuli is named from the Persian 'lazhward', meaning blue, and its uniquely intense colour has been a source of delight for over 6000 years. For many centuries the only known deposits were those at Sar-e-Sang, in a remote mountain valley in Badakhshan, Afghanistan. From here it was exported to the ancient civilizations of Egypt and Sumer (Iraq), and later traded throughout the East and into Europe. These mines are still producing the finest quality lapis lazuli.

Lapis lazuli is a rock composed chiefly of the blue silicate mineral lazurite, together with calcite and brassy coloured pyrite, which are abundant in the poorer quality material (see below). The vivid blue of lazurite is caused by the sulphur that forms an essential part of its chemistry. At Sar-e-Sang the

BELOW Carved bowl of lapis lazuli.

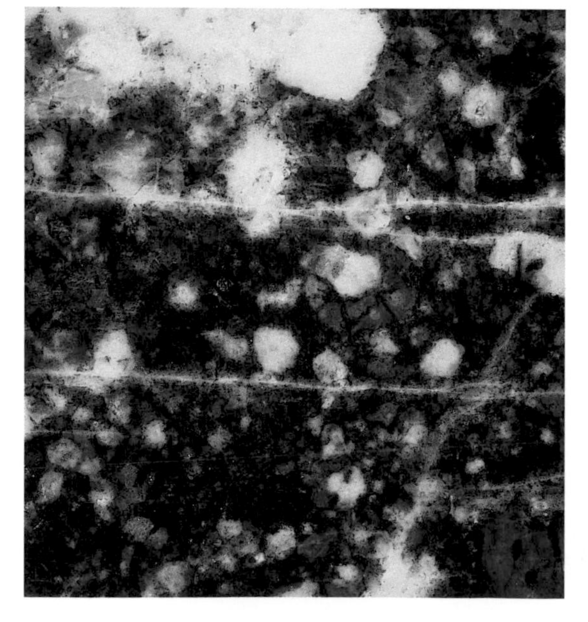

BELOW LEFT Siberian lapis lazuli containing much white calcite.

BELOW RIGHT Beads of lapis lazuli, with rough lapis lazuli, from Afghanistan.

LAPIS LAZULI

CC	a rock composed mainly of lazurite, with minor amounts of calcite and pyrite. Other minerals may be present
H	5.5
SG	2.7–2.9
RI	1.5

lapis lazuli occurs as a zone of lenses and veins within white marble, and grades from deep to pale blue with some violet and greenish tints. Today lapis lazuli is also mined around Slyudyanka in Siberia and in the Ovalle Cordillera in Chile, but material from these sources usually contains more calcite.

Lapis lazuli has always been fashioned as beads and cabochons, carved, or used in inlays and mosaics. In medieval Europe it was crushed to produce the precious pigment ultramarine that was used in many sacred paintings and manuscript illuminations. A substitute was sought for this costly pigment and since 1828 ultramarine has been made artificially.

ABOVE RIGHT Fine dark blue lapis lazuli from Afghanistan.

RIGHT Carved Chinese belt-hook, with rough lapis lazuli from Badakhshan.

Moonstone and Labradorite

The gems featured here are members of the most abundant mineral group in the Earth's crust – the feldspars. Feldspar gems derive their beauty from iridescence, spangled effects or soft subtle colours (see below).

Feldspars are calcium, sodium or potassium aluminium silicates. They very seldom occur in a pure state, but form as two distinct groups – the sodium potassium alkali feldspars and the sodium calcium plagioclase feldspars. Within each group a continuous range of compositions occurs. Quite commonly two of these compositions are intergrown within a single feldspar crystal on a fine or microscopic scale. When light is scattered or reflected by minute intergrowths and suffers interference, a beautiful soft sheen or bright iridescence may be seen, as in moonstone, peristerite and some labradorite. Similar intergrowths, developed on a much larger scale, form the attractive textures of perthite.

The alkali feldspar gems include moonstone, yellow orthoclase and blue-green amazonite (amazonstone). Iron impurities colour yellow orthoclase, while traces of lead and water cause the greenish-blue of amazonite. Labradorite, peristerite and sunstone are plagioclase gems. The bright spangles in sunstone are reflections from tiny, flaky hematite inclusions.

Sri Lanka is the most important source of moonstone, which is mined from pegmatites and the gravels derived from them. The yellow orthoclase from Madagascar, much sunstone, and the fine amazonite from Colorado also occur in pegmatites. Iridescent labradorite is found chiefly in ancient crystalline rocks which formed deep in the crust. Labradorite with the finest iridescence comes from Labrador (Canada) and from Finland where it is known as spectrolite.

ABOVE Moonstone cut as cabochons.

BELOW LEFT Polished slice of labradorite showing iridescence.

BELOW RIGHT Image of a face carved in labradorite.

MOONSTONE AND LABRADORITE

CC	calcium, sodium or potassium aluminium silicate
CS	monoclinic and triclinic
H	6–6.5
SG	2.56–2.76
RI	1.518–1.588
DR	0.006–0.013

Blue John

Blue John is a variety of the mineral fluorite with distinctive purple and colourless or pale yellow banding. It is found at Castleton in Derbyshire, England, where it occurs in variously patterned veins which have been given names such as Twelve Vein and Miller's Vein. Blue John has been used since Roman times, but working reached its peak in the 18th and 19th centuries when Blue John was fashioned into urns, vases and dishes. Blue John is fragile, so is usually bonded with resins for easier working and greater durability.

Rhodochrosite and Rhodonite

FLUORITE

CC	calcium fluoride
CS	cubic
H	4
SG	3.18
RI	1.430

RHODOCHROSITE

CC	manganese carbonate
CS	trigonal
H	4
SG	3.60
RI	1.60–1.80
DR	0.220

RHODONITE

CC	manganese silicate
CS	triclinic
H	6
SG	3.60
RI	1.71–1.73
DR	0.014

LEFT Blue John vase from Derbyshire, England.

ABOVE RIGHT Rhodonite from Kövar, Hungary.

RIGHT Rhodochrosite from Capilitas, Argentina.

Rhodonite and rhodochrosite are pink manganese minerals. Banded rhodochrosite became a popular ornamental stone during the 1930s, after the discovery of fine material in Argentina; this material is sometimes called 'Inca Rose'. In 1974 an important new source was discovered at N'chwaning in Cape Province, South Africa, where both banded rhodochrosite and fine transparent crystals are found. Fine, clear gem-quality crystals, also come from the Sweet Home mine, Colorado. The USA is the main producer of rhodochrosite.

Rhodonite is also a rose pink colour and can be found as plain coloured pieces. More popular, however, and easy to distinguish from rhodochrosite, are pieces that have areas that are black as a result of concentrations of manganese oxides. Rhodonite is found in many countries including Russia, Australia, Sweden,Brazil, Mexico, the USA, Canada, Italy and Britain.

Serpentine

Serpentine is a rock which occurs in a great variety of colours and patterns, sometimes resembling the snakeskin from which it is named. It is soft and easily worked, and has been used for carvings, bowls and inlay work since ancient times. Serpentine consists of the hydrated magnesium silicate minerals chrysotile, antigorite and lizardite. Lizardite was first described from the Lizard in Cornwall, the source of most British decorative serpentine. Bowenite is a translucent green serpentine that is carved extensively in China and is sometimes misleadingly called 'new jade'.

SERPENTINE	
CC	magnesium hydroxysilicate
CS	monoclinic
H	2.5–5
SG	2.60
RI	1.55–1.56

BELOW LEFT Carved tazza comprising three varieties of serpentine, all from the Lizard area of Cornwall, UK.

Ivory and Bone

Ivory is a name given to the teeth or modified teeth (tusks) of elephant, mammoth, walrus, hippopotamus, boar, narwhal and sperm whale. Teeth and tusks are composed mainly of the phosphate mineral hydroxyapatite and organic compounds. Ivories are identified by their structure: for example, elephant and mammoth ivory shows a pattern of intersecting curved lines in cross section, and walrus ivory has a distinctive coarse-grained core.

BELOW MIDDLE Japanese 19th century ivory netsuke showing a female figure.

BELOW RIGHT Elephant ivory carving from Japan.

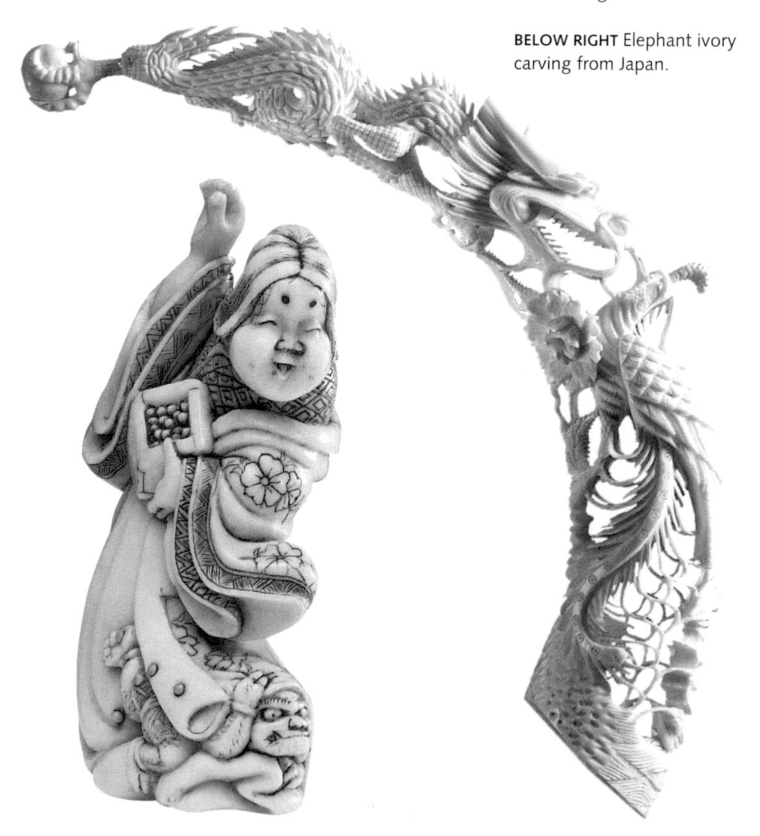

Amber and Jet

Amber is a group name for fossilized tree resins, and includes many resins of differing chemical and physical properties. Trapped animals sometimes provide spectacular proof of its origin. Most commercial amber comes from the Baltic coasts of Russia and Poland, with lesser amounts from the Dominican Republic, Sicily, Myanmar, North America and Mexico. Amber is less dense than most plastic and resin imitations, and will float in salty water. Baltic amber is occasionally washed up on the beaches of eastern Britain.

Jet is fossilized wood, dark brown or black in colour and easily carved and polished. It is found as lens-shaped masses in the Upper Lias shales around the town of Whitby in Yorkshire, England, and also occurs in Spain, France, Germany, Turkey, the USA and Russia. Yorkshire jet has been used since the Bronze Age, and was especially popular in Victorian mourning jewellery. Common imitations include glass, sometimes known as 'Paris jet' or 'French jet', and the rubber derivative vulcanite.

ABOVE Centipede and insects in amber (40–50 million years old).

FAR LEFT Victorian mourning jewellery.

LEFT Faceted beads of amber on a necklace.

Pearl, Coral and Shell

Pearls are produced by some aquatic molluscs, especially oysters and mussels. They are secreted by the soft internal tissues of the animal around an irritant, such as a parasite, and are built up of aragonite (calcium carbonate) layers, known as nacre. The superb iridescent lustre or 'orient' of pearl arises from the interference of light reflected from the boundaries of these thin layers. The finest pearls are produced by marine oysters of the genus *Pinctada*. Pearl mussels are freshwater animals. In Scotland and Ireland, fine pearls have been found in the species *Margaritifera margaritifera*. Some oyster and mussel species are reared in farms to produce cultured pearls. A piece of body tissue, with or without a bead, is inserted into the flesh of the creature, around which the nacre is deposited to form a pearl which is harvested several years later. Cultured pearls have been produced in China for several centuries, but the modern industry was founded in Japan in the late 19th century, and has now spread to China, regions around Australia and the Pacific.

Corals are the skeletons secreted by small sea animals called polyps. Red, pink, white and blue corals are made of calcium carbonate, but black and golden corals are formed of the horny substance conchiolin. In all corals, the skeletal structure is visible as a delicately striped or spotted graining. Mediterranean red and pink coral has been popular for centuries, and was once traded throughout Europe and into India and Arabia. The black and golden corals fished off Hawaii, Australia and the West Indies are much more recent discoveries. Measures are in place to avoid over-exploitation of the natural colonies.

Shells have always delighted us with their elegant shapes and patterns, but the differently coloured layers of helmet shells and giant conch shells are also artistically carved to produce exquisite cameos (left). Mother-of-pearl is the iridescent lining found in many shells, and is especially beautiful in the pearl oyster *Pinctada* (below) and in abalone shells.

ABOVE Red coral partly carved and polished to retain natural branching structure.

LEFT Helmet shell cameo.

BELOW Pearls and oyster shells with pearls strung on silver wires, called a 'Bombay bunch'.

Further Information

FURTHER READING

Color Encyclopedia of Gemstones, 2nd edition J E Arem.Van Nostrand Reinhold, New York, 1987.

Crystal and Gem, R. Symes and R. Harding, DK Eyewitness Handbook series. Dorling Kindersley, London, 1991

Crystals, 2nd edn., G. Cressey and I.F. Mercer. The Natural History Museum, London, 1999.

Diamonds, revised 2nd edn. E. Bruton. NAG Press, London, 1981.

Diamonds revised edn., F. Ward. Gem Book Publishers, Bethesda, 1998.

Diamonds and Precious Stones, P. Voillot. New Horizons series Thames and Hudson, London, 1997.

Emerald and other beryls, J. Sirkankhas. Chilton Press, Radnor, Pennsylvania, 1981.

Emeralds, F. Ward. Gem Book Publishers, Bethesda, 1993.

Gemmologists' Compendium, 7th edn., R. Webster revised by E.A. Jobbins. NAG Press, London, 1998.

Gems, 5th edn., R. Webster revised by P G Read. Butterworths, London, 1994.

Gems and Precious Stones, C. Hall. Apple Press, London, 1993.

Gemstone Enhancement, 2nd edn., K. Nassau. Butterworths, London, 1994.

Gems Made by Man, K. Nassau. Chilton, Radnor, Pennsylvania, 1980.

Gemstones, M. O'Donoghue. Chapman and Hall, London, 1988.

Gemstones, DK Eyewitness Handbook series, 2nd edn, C. Hall. Dorling Kindersley, London, 1994.

Gem Testing, 10th edn. B W Anderson revised by E A Jobbins. Butterworths, London, 1983.

The Nature of Diamonds, G.E. Harlow. Cambridge University Press, New York, 1998.

Rubies and Sapphires, F. Ward. Gem Book Publishers, Bethesda, 1998.

Ruby and Sapphire, R.W. Highes. RWH Publishing, Boulder, Colorado, 1997.

WEB SITES

NB Web site addresses are subject to change.

Collections

http://www.british-museum.ac.uk
[British Museum]

http://www.mnh.si.edu/
[National Museum of Natural History, Smithsonian Institution]

http://www.nhm.ac.uk
[The Natural History Museum]

Education

http://www.scienceweb.org.uk
[resource for children and primary teachers]

Mining

http://www.riotinto.com/ok.html
http://www.mininginformation.com

Professional institutions

http://www.gagtl.ac.uk
[Gemmological Association & Gem Testing Laboratory of Great Britain; professional body for gemmologists in UK]

http://www.gia.org/
[Gemmological Institute of America]

http://www.geolsoc.org.uk
[Geological Society, London, has many useful links to gems, mining websites, earth science and teaching]

http://www.progold.net
[National Association of Goldsmiths, professional body for jewellers in UK]

Trade and auction

http://www.gemnet.co.uk/homepage.html
[gem trade]

http://www.sothebys.com/
[Sotheby's auction house]

http://www.christies.com/
[Christies auction house]

http://www.debeers.com.sg/
[De Beers' Diamond Information Centre]

PICTURE CREDITS

Unless otherwise stated all images are copyright The Natural History Museum, London

British Museum, p. 29, 30 right, 62 left, 64 middle

Christies, p. 31, 33 top

Cobra & Bellamy, p. 32 top, 33 bottom, 54 middle and back cover

De Beer's, p. 9 bottom, 10, 20 bottom left, 21 bottom, 36 top/bottom

Gary Hincks, p. 7 bottom, 18 left, 20 right, 21 top, 35 right

Alan Jobbins, p. 11 top/bottom, 12, 13, 14, 15 top, 17, 22 right, 24, 27 bottom left and right, 28, 47 top right, 65 top, 70 bottom left

Crown © The Royal Collection © 2000 Her Majesty Queen Elizabeth II, p.30 left, 53 left

Victoria and Albert Museum, p.32 bottom, 70 middle, 71 bottom left

Front cover: a selection of faceted gemstones from the collections of The Natural History Museum, London

Inside front cover: Labradorite, from Labrador (see p. 68)

Inside back cover: agate from Idar-Oberstein, Germany (see p. 61)

Back cover: French Art Nouveau brooch set with gems and enamel (see p. 32)

Gem deposits of the world

Gemstones are found worldwide. The principal gem producing countries of the world are listed here together with the gemstones produced in that country. Some areas are famous for a particular gemstone, for example diamonds from South Africa, rubies from Myanmar or emeralds from Colombia, but these countries also have other gemstones. The order of the gemstones gives some idea of their importance today or in historic times, with the most important gemstones from each country at the top of the list.

The symbol * signifies that the gem deposit is historically important but yields little or nothing today.

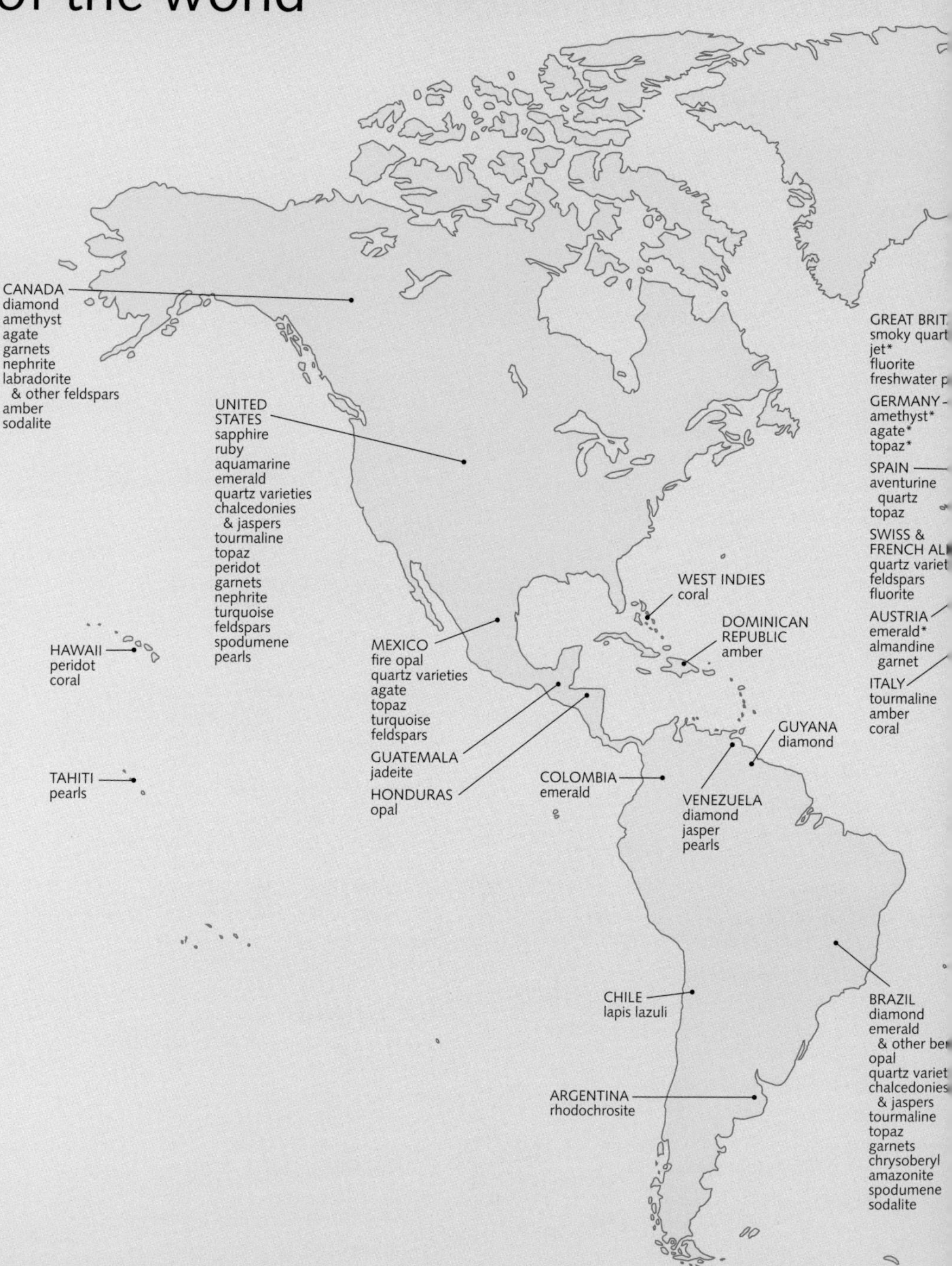

RUSSIA (URALS & SIBERIA)
diamond
emerald & other beryls
quartz varieties
chrysoprase
jasper
tourmaline
topaz
demantoid garnet
alexandrite
nephrite
lapis lazuli
feldspars
rhodonite
malachite
charoite
FINLAND
labradorite
POLAND
chrysoprase*
amber
CZECH REPUBLIC
pyrope garnet*
SLOVAKIA
opal*
NW RUSSIA
amber
diamond
feldspars
freshwater pearls
ROMANIA
chalcedony
amber
rhodonite
CHINA
diamond
ruby
sapphire
aquamarine
amethyst
peridot
nephrite
turquoise
amber
AFGHANISTAN
ruby
aquamarine
tourmaline
spinel
lapis lazuli
spodumene
emerald
JAPAN
quartz varieties
topaz
jadeite
rhodonite
coral
IRAN
turquoise
PAKISTAN
ruby
emerald
aquamarine
grossular garnet
spinel
topaz
peridot
EGYPT
emerald*
jasper
peridot
turquoise*
TAIWAN
nephrite
PERSIAN GULF
pearls
MYANMAR
ruby
sapphire
rock crystal
topaz
tourmaline
peridot
zircon
chrysoberyl
spinel
jadeite
moonstone
scapolite
spodumene
amber
& many rarities
KENYA
ruby
sapphire
aquamarine
amethyst
tourmaline
garnets
feldspars
ZAIRE
diamond
INDONESIA
diamond
SRI LANKA
ruby
sapphire
amethyst
tourmaline
topaz
zircon
garnets
chrysoberyl
spinel
moonstone
enstatite
iolite
& many rarities
TANZANIA
diamond
ruby
sapphire
emerald
aquamarine
tourmaline
garnets
chrysoberyl
feldspars
tanzanite
THAILAND, CAMBODIA, VIETNAM
ruby
sapphire
zircon
almandine garnet
MOZAMBIQUE
smoky quartz
tourmaline
INDIA
diamond
ruby
sapphire
emerald
aquamarine
quartz varieties
chalcedonies
jasper
almandine garnet
chrysoberyl
moonstone
rhodonite
enstatite
star diopside
sodalite
pearls
MADAGASCAR
aquamarine & other beryls
quartz varieties
agate
tourmaline
topaz
garnets
chrysoberyl
feldspars
spodumene
sapphire
ZIMBABWE
emerald
aquamarine
quartz varieties
agate
jasper
tourmaline
topaz
garnets
chrysoberyl
verdite
AUSTRALIA
opal
sapphire
diamond
emerald
chrysoprase
jasper
nephrite
pearls
coral
SOUTH AFRICA
diamond
ruby
emerald
quartz varieties
jasper
tourmaline
peridot
garnets
rhodochrosite
verdite
BOTSWANA
diamond
agate
NEW ZEALAND
nephrite

Index

Bold refers to main references. General subjects refer to general information – for further detail, consult main references under specific stones.

agate 9, 24, 30, 31
alexandrite 50
amazonite 68
amber 12, 28, 34, **71**
amethyst 5, 9, 24, 29, **44-5**
andalusite **56**
aquamarine 26, **40-1**
azurite 9, **65**

benitoite **55**
birefringence 18
bloodstone 60
Blue John **69**
brilliant cut 20, **21**

carats 9, 13
carnelian 29, 60
cat's-eye 50
cat's-eye effect 27, 28, 34
chalcedony 9, 24, **60**, **61**
chrysoberyl 16, 26, 27, 31, **50**, 57
chrysoprase 60
citrine 24, **44-5**
colour 5, 15, 16-18, 20
  treated stones 23-5
composite stones 25-6
coral 34, **72**
cordierite (iolite) **55**
crystals 7, 8, 12, 15, 16, 26
  inclusions 27-8
cutting stones 20-1

diamond 5, 7, 31, 33, 34, **35-7**
  colour 16-17
  famous gems 6, 30
  imitation 19, 22-3, 26, 52
  inclusions 28
  lustre 12
  mining 9-10
dichroic 16
dichroscope 16
diopside 7, **59**
dispersion (fire) 17, 20
durability 6, 14

emerald 6, 16, 23, 24, 31, **40-1**
  imitation 26, 27
  inclusions 27-28
enstatite **59**

fakes 22-28
feldspar 27, 68
fibrolite (sillimanite) **56**

garnet 7, 8, 16, 28, 29, **46-7**
  composite stones 25-6
gravels 10-11, 30

hardness 6, 12-13, 15
hawk's eye 45
hiddenite **57**

imitations 19, 22-3
inclusions 12, 15, 20, 23, 27-8
iolite (cordierite) **55**
ivory 34, **70**

jade 6, 8, 12, 14, 29, **62-3**
  'new jade' 70
  treated 24
jasper **60**, **61**
jet 34, **71**
jewellery 5, 24-5, 29-33

kornerupine **58**
kunzite **57**
kyanite **56**

labradorite 5, 16-17, **68**
lapis lazuli 8, 24, 26, 29-30, **66-7**
lustre 12-13, 20, 26

magnification 14, 27
malachite 9, 16, 34, **65**
map 10-11
mining 9-11, 29-30, 31
Mohs' scale of hardness 6, 13
moonstone 28, 32, **68**

nephrite 6, 12, 62, 63

olivine (peridot) 8, 16, 28, **51**
onyx 60
opal 9, 25, 31, 32, **42-3**
  colours 16-18
  imitation 26, 27
organic gems 34, 70-2

pearls 32, 34, **72**
peridot (olivine) 8, 16, 28, **51**
pleochroism 16
polishing 12-13, 21

quartz 5, 6, 9, 44-5, 60, 61
  composite stones 26
  inclusions 27
  treated 24, 25

rarity 5-6
refraction 12, 15, 18-19, 22, 23
refractometer 19
rhodochrosite **69**
rhodonite 16, **69**
rock crystal 44
rose-cut 21
ruby 8, 16, 23, 30, **38-9**
  Black Prince's 53
  imitation 24, 26

sapphire 8, 11, 14, 16, 30, **38-9**
  imitation 26
  inclusions 27, 28
  treated 23
scapolite 8, **58**
serpentine **70**
shell **72**
sillimanite (fibrolite) **56**
specific gravity (SG) 14
spectroscope 25
spectroscopy 25, 27
sphene (titanite) **59**
spinel 8, 16, 26, 30, **53**
spodumene 8, 57
star stones 34
step-cut 20
synthetic stones 15, 26-7

taaffeite 6, **54**
tanzanite **54**
tiger's-eye 45
titanite (sphene) **59**
topaz 6, 8, 24, 31, **49**
tourmaline 8, 31, **48**
treated stones 15, 23-5
turquoise 9, 12, 16, 24, 29, **64**
  imitation 26

value 5, 13, 20, 27, 30

zircon 6, 8, 19, **52**